Systems Biology 2013

Topical Conference at the 2013 AIChE Annual Meeting: Global Challenges for Engineering a Sustainable Future

San Francisco, California, USA
3-8 November 2013

ISBN: 978-1-63439-027-9

Printed from e-media with permission by:

Curran Associates, Inc.
57 Morehouse Lane
Red Hook, NY 12571

Some format issues inherent in the e-media version may also appear in this print version.

Printed by Curran Associates, Inc. (2014)

For permission requests, please contact AIChE
at the address below.

AIChE
3 Park Avenue
New York, NY 10016-5991

Phone: (203) 702-7660
Fax: (203) 775-5177

www.aiche.org

Additional copies of this publication are available from:

Curran Associates, Inc.
57 Morehouse Lane
Red Hook, NY 12571 USA
Phone: 845-758-0400
Fax: 845-758-2634
Email: curran@proceedings.com
Web: www.proceedings.com

TABLE OF CONTENTS

(72a) Engineering Biological Systems: The Increasing Role of Design in a Time of Paradigm Shifting Technology

Gill, R. T., University of Colorado Boulder

Originally presented on: 11/4/2013 8:30:00 - 9:30:00

Technologies for synthesizing and sequencing DNA have advanced faster than Moore's law, resulting in completely new ways of thinking about how to go about engineering microbial genomes. For example, the first complete microbial genome was synthesized and successfully transplanted into a host chassis in 2010 at a cost of ~$10-100 million. In just 3 years, synthesis technology has advanced to the point that it is now expected that sufficient DNA to construct full genomes will be available at costs 100,000x lower and in turn-around times of weeks. Sequencing has advanced even further, with the latest technologies proposing full-genome microbial sequencing for under $100. Along with improved methods for screening, sorting, and characterizing cells, the throughput of the biological design-build-test cycle has increased by several orders of magnitude. Such capabilities require a shift in thinking at the most basic level about how to pursue microbial systems engineering, with much recent inspiration in this arena drawn from computer science and engineering, and specifically throughout the rapidly expanding field of synthetic biology. While synthetic biology approaches have focused primarily on a ground-up approach to cellular engineering, metabolic engineering and systems biology have emphasized engineering strategies that emphasize the performance of the overall system as a starting point. Future efforts will bridge these two paradigms, with the initial step focusing on rigorous design efforts that not only attempt to connect well characterized biological parts to overall performance through systems modeling but also on design search algorithms that guide design-build-test-learn cycles in a manner that increases understanding and furthers the engineering effort. Our efforts are focused on developing technologies and approaches that enable this vision. This presentation will describe several of our efforts in this area, as well as provide recent data on application areas we are pursuing including the biological production of chemicals and fuels.

(72b) Rewriting Pathways to Life

Liao, J. C., University of California Los Angeles

Originally presented on: 11/4/2013 9:30:00 - 10:30:00

In the first century of biochemical research, the focus has been on discovery and understanding of mechanisms for cellular reactions, signaling, and regulation. To date, the pathways and genes involved in cellular metabolism have largely been mapped out, along with reaction mechanisms in sufficient details. The time is ripe for re-designing metabolic networks for a specific purpose, say, replacing petroleum, which serves as the raw material for the majority of fuels and chemicals used today. However, the primary metabolic pathways, such as glycolysis and the Calvin cycle, are evolved for the purpose of cell growth and survival, rather than for biochemical productions. Glycolysis, or its variations, is a fundamental metabolic pathway in life that exists in almost all organisms to decompose external or intracellular sugars. The pathway proceeds through partial oxidation and splitting of sugars to pyruvate, which in turn is decarboxylated to produce acetyl-coenzyme A (CoA) for various biosynthetic purposes. The decarboxylation of pyruvate loses a carbon equivalent, and limits the theoretical carbon yield to only two moles of two-carbon (C2) metabolites per mole of hexose. This native route is a major source of carbon loss in biorefining and microbial carbon metabolism. Here we design and construct a non-oxidative, cyclic pathway that allows the production of stoichiometric amounts of C2 metabolites from hexose, pentose, and triose phosphates without carbon loss. This pathway, termed Non-Oxidative Glycolysis (NOG) enables complete carbon conservation in sugar catabolism to acetyl-CoA, and can be used in conjunction with CO_2 fixation and other one-carbon (C1) assimilation pathways to achieve 100% carbon yield to desirable fuels and chemicals. In this talk, we will discuss how NOG was designed and the role of Systems Biology in this effort.

(141a) Use of Computational Biology and Gene Expression Array Analysis to Understand Human Brain Evolution

Grossman, L. I., Wayne State University
Chugani, H., Wayne State University
Kuzawa, C., Northwestern University
Lipovich, L., Wayne State University
McGowen, M., Wayne State University
Sherwood, C., George Washington University
Sterner, K., Wayne State University
Wildman, D. E., Wayne State University School of Medicine
Originally presented on: 11/4/2013 12:30:00 - 12:48:00

What makes us human? Considering how minor the changes in DNA sequence are between ourselves and our nearest primate relatives, this is a question of abiding interest. We have focused on the brain to address this question in a series of evolutionary genomic studies. The goal is to identify both biological processes and individual genes that have been important during human evolution. We first examined lineages (cetaceans and proboscideans) in addition to human that produced large brained species (dolphins and whales, and elephants). We utilized dN/dS, the ratio of nonsynonymous (amino acid replacing) nucleotide substitutions per nonsynonomous site to synonymous (amino acid unchanging) substitutions per synonymous site, as a measure of positive natural selection and identified the biological categories in which positively selected genes were found. Humans, dolphins, and elephants all showed genes expressed in mitochondria that are significantly overrepresented among positively selected genes. This finding underlines the evolutionarily parallel molecular trajectories of groups possessing a large brain, among the most energy consuming tissues. We then studied age-specific gene expression in human neocortex, which is characterized by protracted developmental intervals of synaptogenesis and myelination that allow for an extended period of learning, taken from the nondiseased margins of surgically resected tissue. We found 8 nonmessenger, long nonprotein-coding ribonucleic acid (lncRNA) molecules that are differentially expressed in an age-dependent way, 337 transcripts with greater inter-individual variance among children than among adults – many annotated to the immune system – and 40 transcripts with significant age-related trajectories in expression. These findings present some of the genetic underpinnings of the extended period of human cortical development with peak brain glucose metabolism between ages 4 and 10 years. (Supported by NSF BCS-0827546)

(141b) Comparative Analysis of the Cerebellar Metabolic Physiology Between Sexes in a Mouse Model of Adult Onset Hypothyroidism Using GC-MS Metabolomics

Maga-Nteve, C., University of Patras
Vasilopoulou, C., Foundation for Research and Technology-Hellas (FORTH)
Constantinou, C., Foundation for Research and Technology - Hellas (FORTH)
Margarity, M.,
Klapa, M. I., Foundation for Research and Technology-Hellas (FORTH)
Originally presented on: 11/4/2013 12:48:00 - 13:06:00

In the systems biology era, the high-throughput "omic" technologies have enabled the holistic analysis of the various molecular levels of cellular function through the simultaneous measurement of hundreds to thousands of relevant molecular quantities. Metabolomics refers to the quantification and analysis of the (relative) concentration profile of the free small metabolites, which provides a fingerprint of the metabolic state of a biological system. Among the advantages of the metabolomic analysis is that it can be easily used to monitor transient metabolic conditions without requiring extensive knowledge of the structure and regulation of the investigated metabolic networks. This characteristic is especially advantageous for the analysis of brain metabolism, considering the anatomical, morphological and phenotypic complexity of this organ and our current shortages in understanding its metabolic mechanisms. For the effect of adult onset hypothyroidism (AOH) on brain metabolism in particular, the current knowledge remains fragmented, concerning different experimental setups and recovered from various brain regions. Therefore, a holistic view of the metabolic response of particular brain regions to AOH is expected to significantly enhance the current knowledge about the disease. In this context, we used metabolomics to study the metabolic physiology of mouse cerebellum in a prolonged AOH model. This was the first metabolomic analysis of brain tissue in a prolonged AOH mouse model [1]. We extended our analysis to both sexes to investigate potential differences of the effect of prolonged AOH on the cerebellar metabolic physiology between male and female.

The prolonged AOH was induced by a 64-day treatment with 1% potassium perchlorate in the drinking water of Balb/cJ mice, as described in our published study [1]. The raw metabolic profiles were normalized and appropriately filtered based on the methodology of Kanani and Klapa [2]. The normalized metabolic profiles were analyzed using the open-source TM4/MeV software (www.tm4.org/mev) for the multivariate statistical analysis of "omic" data. The acquired results were interpreted in the context of an appropriately reconstructed metabolic network for the mouse cerebellum based on the metabolic databases, KEGG (www.kegg.com) and Expasy (enzyme.expasy.org), and a plethora of information mined from the literature. The analysis of the acquired metabolic profiles provided strong evidence that the mammalian cerebellum is metabolically responsive to prolonged AOH [1]. Specifically, it indicated a general decline in the metabolic activity of cerebellum in the hypothyroid compared to the euthyroid animals, with the observed effect for the particular AOH mouse model being stronger in the male compared to the female. Our results enhanced the currently available "omic" dataset for the brain physiology under prolonged AOH and indicated directions for further research to improve our understanding of AOH effect on brain metabolism. Integration with other omic, in particular proteomic, data would be of great value.

References

1. Constantinou C., Chrysanthopoulos P., Margarity M. and Klapa MI. (2011) GC-MS metabolic analysis reveals significant alternations in cerebellar metabolic physiology in a mouse model of adult onset hypothyroidism. *J. Proteome Res.* **10**:869-879 (Epub 2010 Dec 8)
2. Kanani H., Klapa MI. (2007) Data Correction Strategy for metabolomic analysis using Gas Chromatography-Mass Spectrometry. *Metab. Eng.* **9**: 39-51.

(141c) Nonlinear Dimensionality Reduction for Cellular Phenotyping From High Throughput Mass Cytometry Data

Shekhar, K.,
Brodin, P.,
Chakraborty, A. K., Massachusetts Institute of Technology
Davis, M. M., Stanford University, Howard Hughes Medical Institute
Originally presented on: 11/4/2013 13:06:00 - 13:24:00

The advent of mass cytometry has enabled simultaneous, high-throughput measurements of ~40 proteins at the single cell level. Such unprecedented data makes possible, for the first time, resolution of cellular phenotypic diversity under normal conditions, and also systematic classification of alterations therein resulting from pathologies. Current techniques to interrogate cellular diversity use unsupervised clustering to classify individual cells into groups and produce visualizations in the form of a graph structure describing the relatedness of different cell types. Often, a major limitation in these methods is that number of clusters needs to be pre-specified by the user, and furthermore, the final visualization produced by the algorithm relies heavily on prior knowledge of existing cell types. We describe an alternative approach based on non-linear dimensionality reduction that overcomes these limitations, and demonstrate its advantages by applying it on two datasets. Our technique produces a two-dimensional embedding of cells based on their relative expression patterns and a simple kernel density estimator identifies unique subpopulations with significantly different expression patterns. We first analyzed published human bone marrow data and show that our method produces results consistent with the existing state-of-the-art clustering-based analysis tool. Furthermore, our unbiased density-based detection of cell types also identifies subsets not described under the existing taxonomy. Application of our method on CD8 T cells extracted from wild-type Black-6 mice of different ages, while demonstrating the stereotypy of cellular development in genetically identical mice, also identifies a distinct subpopulation of naïve cells that vary as a function of age.

(141d) A Quantitative Systems Perspective On Cancer, Immunoevasion, and Somatic Evolution
Klinke, D. J., West Virginia University
Originally presented on: 11/4/2013 13:24:00 - 13:42:00

Following from the clinical challenges associated with molecular targeted therapies, there is increasing realization that one should view cancer from an evolutionary perspective [1]. Evolutionary processes, like cancer, have three common traits: heterogeneity, dynamics, and a selective fitness landscape. Mutagens randomly alter the genome of host cells creating a population of cells that contain different somatic mutations. This genomic rearrangement perturbs cellular homeostasis through changing how cells interact with their tissue microenvironment. Control structures are encoded within cells and organ systems, such as innate and adaptive immunity, to restore cellular homeostasis. These control structures shape the selective fitness landscape and determine whether cells that harbor particular somatic mutations are retained or eliminated from a cell population. While next-generation sequencing has revealed the complexity and heterogeneity of oncogenic transformation, identifying how control structures regulate tissue homeostasis in health and disease is a central challenge in translating our knowledge of biological components into viable therapeutic strategies. In this talk, I will summarize our recent efforts to test postulated intra- and intercellular control structures against the available data using simulation and identify novel mechanisms of cellular cross-talk that cancer cells use to alter the fitness landscape.

References:
1. Klinke DJ. "An evolutionary perspective on anti-tumor immunity" Front Oncol. 2012;2:202.

(141e) The Human Platelet Reconstruction: Understanding Mechanistic Outcomes From Proteomics, Metabolomics, and Fluxomics

Thomas, A., University of California - San Diego
Jamshidi, N.,
Palsson, B., University of California, San Diego
Originally presented on: 11/4/2013 13:42:00 - 14:00:00

Systems biology catalogues knowledge in a framework which enables mechanistic insights to the causes and effects of resulting biological phenotypes, allowing for the contextualization of datasets which subsequently promote the predictive capabilities of biological models. Constraints based methods quantitatively model biochemical reactions while annotating associations which influence the intracellular state of fluxes through pathways, and have been applied to discover antibiotics, propose metabolic engineering strategies, and understand principles in basic biology. With the reconstruction of human metabolism [1, 2], constraints based methods have been applied to cell-specific models, however, properly modeling human metabolism generally requires a metabolically active cell-type and the annotation of complex, altruistic, non-metabolic processes such as signaling, transcriptional regulation, and cell-to-cell community interactions. Due to these complexities, modeling simpler, metabolically-focused human cell types presents an opportunity to build mass balance models that can incrementally incorporate other modeling paradigms. Blood platelets represent one such system. Platelets are enucleated, have simple cellular fates, and are metabolically active. Signaling pathways associated with thrombosis are shallow with known mechanistic outcomes. In addition, reconstructing the metabolism of these blood cells offer opportunities for applications to improve quality of blood transfusions. Therefore, since these processes have not been characterized at a cell-wide-level, a biochemical reaction network of platelet metabolism was constructed using evidence from 33 proteomic datasets and over 300 literature references. The flux balance analysis model was built using a modified model building algorithm [3] with map-driven gap filling. Functional verification of the in vivo energy pathways was performed with isotopomer labeling data [4]. The data were mapped onto the model employing a parsimonious sampling algorithm which samples possible media uptake and secretion choices to define a minimal media for in vivo energy sources. The effects of SNPs, drugs, and phosphorylation events on the biochemical state of the platelet were enumerated. In addition to characterizing intracellular flux states of platelets in vivo, this study puts forth a pipeline and set of computational methods to reconstruct and study human cells in the context of proteomics and isotopomer data.

[1] Duarte, Natalie C., et al. "Global reconstruction of the human metabolic network based on genomic and bibliomic data." Proceedings of the National Academy of Sciences 104.6 (2007): 1777-1782.

[2] Thiele, Ines, et al. "A community-driven global reconstruction of human metabolism." Nature biotechnology (2013).

[3] Livnat Jerby, Tomer Shlomi, and Eytan Ruppin. "Computational reconstruction of tissue-specific metabolic models: application to human liver metabolism."Molecular systems biology 6.1 (2010).

[4] Guppy, Michael, et al. "Fuel choices by human platelets in human plasma."European Journal of Biochemistry 244.1 (1997): 161-167.

(141f) Reverse Engineering Validation Using a Benchmark Synthetic Gene Circuit in Human Cells

Kang, T., University of Texas at Dallas
White, J., University of Texas at Dallas
Xie, Z., Tsinghua University
Benenson, Y., Harvard University
Sontag, E.,
Bleris, L., University of Texas at Dallas
Originally presented on: 11/4/2013 14:00:00 - 14:18:00

Multi-component biological networks are often understood incompletely, in large part due to the lack of reliable and robust methodologies for network reverse engineering and characterization. As a consequence, developing automated and rigorously validated methodologies for unraveling the complexity of biomolecular networks in human cells remains a central challenge to life scientists and engineers. Today, when it comes to experimental and analytical requirements, there exists a great deal of diversity in reverse engineering methods, which renders the independent validation and comparison of their predictive capabilities difficult[1, 2, 3, 4]. While these methods have demonstrated successful network reconstructions on its own, the lack of unifying standards and procedures for validation, along with the high degree of expertise required for computational algorithms, can be regarded as one of the major obstacles which prevents the widespread use of network inference methods. To address this issue, a community-wide effort, DREAM (Dialogue for Reverse Engineering Assessments and Methods), has been initiated to facilitate discussion and refine existing methodologies, resulting in valuable insights about relationships between algorithm performance and experimental parameters[5, 6, 7, 8, 9].

In this study[10] we describe an experimental and theoretical platform customized for the development and verification of reverse engineering and pathway characterization algorithms in mammalian cells. Specifically, we stably integrate a synthetic gene network in human kidney cells and use it as a benchmark for an in vitro platform for reverse engineering verification. The network, which resembles a natural network topology, is orthogonal to endogenous cellular signaling pathway. It combines a set of regulatory interactions consisting of transcriptional and post-transcriptional regulatory elements, where the output fluorescent proteins are subject to three different modes of control: no activation, single source of activation, and combination of activation and repression. As our baseline reverse engineering method we use an approach based on Modular Response Analysis (MRA), where we take experimentally measured steady-state responses following near-linear perturbation of each modular component of the benchmark system[11, 12, 13]. After performing successive perturbations to each modular components of the network, the pre- and post-perturbation steady states were then used to predict the network structure. To capture the system's response after each perturbation, we chose to utilize protein and mRNA as the two representative species. We compare the results of the algorithm against the intended network structure, and use these results to determine the appropriate statistical procedure for each measurement in order to increase the prediction confidence.

1. Bansal, M., Belcastro, V., Ambesi-Impiombato, A., and di Bernardo, D. (2007) How to infer gene networks from expression profiles *Mol. Syst. Biol. 3*, 78.

2. Sprinzak, D., and Elowitz, M. B. (2005) Reconstruction of genetic circuits *Nature. 438*, 443-448.

3. Albert, R., Dasgupta, B., and Sontag, E. (2010) Inference of signal transduction networks from double causal evidence. *Methods Mol. Biol. 673*, 239-251.

4. Hendrickx, D. M., Hendriks, M. M., Eilers, P. H., Smilde, A. K., and Hoefsloot, H. C. (2011) Reverse engineering of metabolic networks, a critical assessment *Mol. Biosyst. 7*, 511-520.

5. Stolovitzky, G., Monroe, D., and Califano, A. (2007) Dialogue on reverse-engineering assessment and methods: The DREAM of high-throughput pathway inference *Ann. N. Y. Acad. Sci. 1115*, 1-22.

6. Stolovitzky, G., Prill, R. J., and Califano, A. (2009) Lessons from the DREAM2 challenges. *Ann. N. Y. Acad. Sci. 1158*, 159-195.

7. Marbach, D., Prill, R. J., Schaffter, T., Mattiussi, C., Floreano, D., and Stolovitzky, G. (2010) Revealing strengths and weaknesses of methods for gene network inference *Proc. Natl. Acad. Sci. U. S. A. 107*, 6286-6291.

8. Marbach, D., Costello, J. C., Kuffner, R., Vega, N. M., Prill, R. J., Camacho, D. M., Allison, K. R., Kellis, M., Collins, J. J., and Stolovitzky, G. (2012) Wisdom of crowds for robust gene network inference. *Nat Meth. 9*, 796-804.

9. Prill, R. J., Saez-Rodriguez, J., Alexopoulos, L. G., Sorger, P. K., and Stolovitzky, G. (2011) Crowdsourcing network inference: The DREAM predictive signaling network challenge *Sci. Signal. 4*, mr7.

10. Kang, T., White, J. T., Xie, Z., Benenson, Y., Sontag, E., and Bleris, L. (2013) Reverse engineering validation using a benchmark synthetic gene circuit in human cells. *ACS Synth Biol.* http://pubs.acs.org/doi/abs/10.1021/sb300093y

11. Kholodenko, B., Yaffe, M. B., and Kolch, W. (2012) Computational approaches for analyzing information flow in biological networks. *Science signaling. 5*, re1.

12. Kholodenko, B. N. (2007) Untangling the signalling wires *Nat. Cell Biol. 9*, 247 <last_page> 249.

13. Sontag, E. D. (2008) Network reconstruction based on steady-state data. *Essays in biochemistry. 45*, 161-76.

(141g) A Naive Bayesian Classifier of Gram Stain Phenotypes From Genotype Functional Roles
Colasanti, R.,
Originally presented on: 11/4/2013 14:18:00 - 14:36:00

A naive Bayesian classifier of Gram stain phenotypes from genotype functional roles
Ricardo L Colasanti[1,*], Janaka N Edirisinghe [1], Christopher S Henry[1]
1 Mathematics and Computer Science Division, Argonne National Laboratory, 9700 South Cass Avenue, Argonne, IL 60439 USA
∗ E-mail: ric.colasanti@gmail.com
Abstract

The recent advances in genome sequencing mean that we now have access to such vast quantities of data on the genetic codes of bacteria that it is increasingly difficult to analyse them, and deduce a hypothesis [3]. This linked with the increased scope of 'Omics' technologies: genomics, transcriptomics, proteomics, metabolomics, epigenomics and metagenomics [4] means that biological hypothesis formation must become more automated. Metagenomics is an interesting example. Much of modern microbial ecology has moved from the Petri dish and growth media as a means of identification, to the sequencer. In fact the term Ecosystomics has regretfully been coined [1]. And microbial ecology is at the heart of modern science with the human microbiome project [5] now an important research area in human health.

We are exploring the use of naive Bayesian classifiers as a small step in the quest for automated hypothesis formation. We set out to predict the phenotypic behaviour of a bacteria from a list of annotated biological functions. As a test of our approach, we selected the identification of Gram stain classification of a bacteria. Because the gram stain is a marker for peptidoglycan, which is present in the thick-layered walls of Gram +ve bacteria and because the cell wall defines the biochemistry of the cell, we expect to see clear functional signatures of Gram stain classification in genome annotations.

There has been previous work with machine learning of genomes [2]; our work differs in that we are utilized annotated functions as predictors, instead of protein sequences. There has been a large amount of work done on encoding of biological functions, giving descriptions of bacteria in terms of the roles that their proteins perform [6]. The use of functional roles allows us to conflate numerous genome sequences together as pre-classified classes. Genomic databases, such as SEED or the DOE Knowledgebase, contain a large set of functional roles and the sets of genomes that give rise to them. Some functional roles are common to all bacteria, some are specific to certain types. We are using machine learning methods to find underlying patterns in the presence or absense of functional roles with the aim of being able classify bacteria based on their annotations.

With bacterial Gram phenotype classification, we have taken the same machine learning approach as that taken by document filters. Each bacteria is represented as a document containing functional role words. We have data of bacteria classified as Gram+ve and Gram−ve. We used the relative frequency of each functional role in Gram+ve and Gram−ve bacteria to classify unseen bacteria.

The classifier was trained by first reading through the training set and creating a list of functional role attributes. The number of times each functional role occurs in a Gram+ve and Gram−ve bacteria is counted. This is used to calculate a conditional probability of each functional role given either Gram+ve or Gram−ve. These probabilities are used to calculate the probability of an unseen bacteria being Gram+ve or Gram−ve. This is accomplished by a Bayesian combined conditional probability of the functional roles of the unseen bacteria to produce a relative probability of that bacteria being Gram+ve or Gram−ve

The results indicate the power of applying naive predictors to automatically classify genomes with any set of phenotypes for which a diverse training set of genomes is available. The classifier was trained on 86 known Gram +ve bacteria and 252 Gram −ve bacteria . After training, it was tested on 112 unseen bacteria; the classifier exhibited an average accuracy of 0.97 and a balanced accuracy measurement of 0.96. On average the classifier only misclassified 3 bacteria on any single test. An examination of the list of all the bacteria that were misclassified provided a couple of interesting insights. The genus Deinococcus is actually a very difficult bacteria to stain with crystal violet and as such is difficult to classify within the Gram classification. The bacteria Syntrophomonas wolfei was the only Gram−ve bacteria that was misclassified as a Gram−ve. Interestingly this bacteria is unique in having a multilayered cell wall and its lack of internal membrane-bound organelles. This is perhaps why the classifier identifies it as a Gram+ve bacteria, because as we have noted the cell wall defines the biochemistry of the cell and the biochemistry is a function of the collective functional roles of the enzymes. The biochemistry of the stout walled Gram−ve is probably closer to that of the thick walled Gram+ve bacteria. The reasons for the other misclassified bacteria are more difficult to ascertain, but may be due to a less than perfect annotation of their genome. However, we emphasize how our naive classifier, built based only on an input training set of preclassified genomes (and a database of consistent genome annotations) exposes new biology for deeper analysis.

The tools used to produce our classifier from a training set, and the tools used to apply our classifier to any new genome based only on its genome sequence have been built into the DOE Systems Biology Knowledgebase, enabling their application for new studies by the scientific community. We will highlight and demonstrate these tools in the context of the analysis of our new Gram stain classifier. References

1. Anthony M Poole, Daniel B Stouffer, and Jason M Tylianakis. ecosystomics: ecology by sequencer. Trends in Ecology and Evolution, 27(6):309, 2012.

2. Margaret A Shipp, Ken N Ross, Pablo Tamayo, Andrew P Weng, Jeffery L Kutok, Ricardo CT Aguiar, Michelle Gaasenbeek, Michael Angelo, Michael Reich, Geraldine S Pinkus, et al. Diffuse large b-cell lymphoma outcome prediction by gene-expression profiling and supervised machine learning. Nature medicine, 8(1):68–74, 2002.

3. A.L. Tarca, V.J. Carey, X. Chen, R. Romero, and S. Drăghici. Machine learning and its applications to biology. PLoS computational biology, 3(6):e116, 2007.

4. Wouter G Touw, Jumamurat R Bayjanov, Lex Overmars, Lennart Backus, Jos Boekhorst, Michiel Wels, and Sacha AFT van Hijum. Data mining in the life sciences with random forest: a walk in the park or lost in the jungle? Briefings in bioinformatics, 2012.

5. Peter J Turnbaugh, Ruth E Ley, Micah Hamady, Claire M Fraser-Liggett, Rob Knight, and Jeffrey I Gordon. The human microbiome project. Nature, 449(7164):804–810, 2007.

6. T.Y. Wong, L.A. Preston, and N.L. Schiller. Alginate lyase: review of major sources and enzyme characteristics, structure-function analysis, biological roles, and applications. Annual Reviews in Microbiology, 54(1):289–340, 2000.

(141h) Reduction Techniques and Hybrid Modelling Approaches for Cell Population Heterogeneity

Stamatakis, M., University of Delaware
Originally presented on: 11/4/2013 14:36:00 - 14:54:00

Latest research on biological systems is steadily shifting from isolated single cells to entire cell populations. The latter are inherently heterogeneous, in the sense that the properties of the single cells that make up the population are typically not the same. To model such populations, three approaches that have been widely used are (i) the cell population balance, (ii) ensemble and (iii) continuum modelling frameworks. Each of these approaches focuses on different aspects of the processes of growth, division and intracellular reaction occurrence and can provide information at different levels of detail, with the corresponding computational expense. Continuum models neglect heterogeneity by assuming lumped biomasses, capture the dynamics of bulk intracellular concentrations and are easy to simulate. Ensemble models account for heterogeneity due to different initial conditions or kinetic constants, and are more computationally expensive. Finally, the cell population balance (CPB) approach captures cell growth, intracellular reactions, division and the partitioning of the intracellular contents in detail, but can quickly become intractable, as the number of biochemical species taken into consideration increases.

In this work, we show how one can reduce the CPB to hybrid modelling approaches that simulate cell population heterogeneity at a much lower computational expense. Under certain assumptions, these hybrid models are exact alternatives of the cell population balance; no approximation is involved.[1] We further demonstrate the use of these hybrid models to efficiently simulate cell population heterogeneity in a genetic network of a single gene with feedback.[2] Starting from a 4-species model we use singular perturbation analysis to derive a single equation for the intracellular protein concentration. We subsequently incorporate this equation to a hybrid model consisting of a CPB for the cell volume, and a continuum equation for the protein concentration. We finally compare the results obtained with the hybrid model with those of the full CPB, demonstrating the accuracy and computational efficiency of the hybrid methodology.

References

1. Stamatakis, M., *Cell population balance, ensemble and continuum modeling frameworks: Conditional equivalence and hybrid approaches.* Chemical Engineering Science, 2010. **65**(2): p. 1008-1015.
2. Stamatakis, M., *Cell Population Balance and Hybrid Modeling of Population Dynamics for a Single Gene with Feedback.* Computers & Chemical Engineering, 2013. **53**: p. 25-34.

(190a) IMPACT Of Intracellular Crowding Conditions On Reaction Rates

Angeles-Martinez, L., University of Manchester
Theodoropoulos, C., University of Manchester
Originally presented on: 11/4/2013 15:15:00 - 15:33:00

Several chemicals are produced by microbial fermentation, for example the succinic acid that can be synthesized from glycerol by sundry microorganisms, among them *Actinobacillus succionogenes* which we have been studying in the past few years [1,2].

However, the synthesis of a metabolite of interest is in general accompanied by the formation of by-products. The simulation and analysis of the factors that affect the metabolism allow its further manipulation in order to improve the production of the target metabolite and minimize the by-products one.

The environmental conditions play an important role on reactions, and the cell contains a complex, heterogeneous, and crowded media, where the volume occupied by the molecules represent about 40% of the total volume. Moreover, the number of enzymes and metabolites in a fraction of the cellular space can be large [3].

Due to the impossibility that two molecules occupy the same at the same time, the intracellular crowding conditions affect the reactions rate in two opposite ways: 1) decreasing the diffusion of the molecules, and 2) increasing their thermodynamic activity [4].

Several methodologies have been proposed to simulate reaction-diffusion systems, among them is kinetic Monte Carlo (kMC), where each molecules is tracked during its journey, although this respects the impenetrability of the molecules it become a computationally expensive task for long times simulation and/or large number of molecules.

On the other hand Lattice Boltzmann Method (LBM) can efficiently handle a large number of molecules, nevertheless their volume is neglected, and therefore also the steric repulsion.

The aim of this work is to propose a new LBM-based methodology (called cLBM) that incorporates the macromolecular crowding effects on the simulation of reaction-diffusion systems as those carried out inside the cell.

In order to validate this new methodology the results are compared with those obtained from an in-house developed lattice kMC algorithm. We use a small system of enzymatic reactions in presence of inert molecules at different concentrations as case study. Although discrepancies are found between both methodologies, cLBM provides the basis for the further simulation of metabolic networks under crowding conditions.

References

1. Vlysidis, A., Binns, M., Webb, C., and Theodoropoulos, C. 2011. Glycerol utilisation for the production of chemicals: conversion to succinic acid, a combined experimental and computational study. Biochemical Engineering Journal, 58-59: 1-11.
2. Angeles-Martinez, L., and Theodoropoulos, C. 2012. Thermodymanic constraints in Flux Balance Analysis. Manuscript in preparation.
3. Goodsell, D.S. 1991. Inside a living cell. Trends in Biochemical Sciences 16, 203-206.
4. Chebotareva, N.A., Kurganov, B.I., and Livanova, N.B. 2004. Biochemical effects of molecular crowding. Biochemistry (Moscow) 69:1239-1251.

(190b) Global Protein Motions in the Allosteric Regulation of CRP/FNR Family Transcription Factors

Rodgers, T. L., The University of Manchester
Townsend, P., Durham University
Burnell, D., Durham University
McLeish, T., Durham University
Pohl, E., Durham University
Wilson, M., Durham University
Cann, M., Durham University
Originally presented on: 11/4/2013 15:33:00 - 15:51:00

Allostery is a fundamental process by which ligand binding to a protein alters its activity at a distinct site. There is growing evidence that allosteric co-operativity can be communicated by modulation of protein dynamics. We provide a foundational theory for how allostery can occur as a function of low frequency dynamics without a change in protein structure. We have generated coarse-grained models that describe the protein backbone motions of the homodimeric CRP/FNR family transcription factors, Catabolite Activated Protein (CAP) and GlxR. We demonstrate that binding the first molecule of cAMP ligand modulates the global normal modes resulting in negative co-operativity for binding the second cAMP ligand without a change in mean structure. The theory makes key experimental predictions and these are validated through an analysis of variant proteins by a combination of structural biology and isothermal calorimetry. A quantitative description of allostery as a free energy landscape revealed a protein 'design space' that identified the key inter- and intramolecular regulatory parameters that contribute to CRP/FNR family allostery.

(190c) Analysis of Complex Biochemical Reaction Network Behavior: Recent Advances in Species-Reaction Graph Theory

Knight, D., Purdue University
Shinar, G.,
Feinberg, M., Ohio State University
Originally presented on: 11/4/2013 15:51:00 - 16:09:00

Analysis of Complex Biochemical Reaction Network Behavior: Recent Advances in Species-Reaction Graph Theory

Daniel Knight[1], G. Shinar[3], Martin Feinberg[1,2]

[1]Department of Chemical and Biomolecular Engineering and the [2]Department of Mathematics, Ohio State University, Columbus, Ohio
[3]Javelin Medical Ltd., 4 Pekeris St., Rehovot 76702, Israel

When studying biological systems, predicting their dynamical properties is more readily inferred from knowledge of only the reaction network, as detailed kinetic information is often not available with confidence. Our goal is to be able to make powerful statements about what dynamic behavior a (potentially very complex) system might admit based only on the network structure, provided the kinetics falls within a very general and natural class. In particular, we wish to distinguish in a precise way between networks that might provide the basis for a bistable switch, and networks – even very intricate ones – that cannot. The foundation for these results is the Species-Reaction, or SR graph, which is an intuitive method of displaying a reaction network not dissimilar from classical biological network representations. Improvements have been made to recent results[1,2], drawing a stronger connection between the properties of a network's SR graph and its capacity to exhibit interesting dynamical features, such as multiple steady states. This new theorem will be given, and a few example networks will be discussed.

1. G. Shinar and M. Feinberg. "*Concordant Chemical Reaction Networks.*" Math. Biosci. 240.2 (2012): 92-113.

2. G. Shinar and M. Feinberg. "*Concordant Chemical Reaction Networks and the Species-Reaction Graph.*" Math. Biosci. 241 (2013): 1-23.

(190d) Analysis and Simulation of Multi-Species Dynamic Flux Balance Analysis Models With Obligatory Interactions

Hoeffner, K., Massachusetts Institute of Technology
Jimenez, E.,
Barton, P. I., Massachusetts Institute of Technology
Originally presented on: 11/4/2013 16:09:00 - 16:27:00

Most microorganisms in natural environments do not live in isolation, but exist as part of complex, dynamically changing, microbial consortia. Hence, synthesis of artificial biological process systems based on microbial consortia is a very promising approach. In addition, monocultures optimized for high metabolic production in a sterile laboratory environment are often not economical at production scale. In contrast, naturally occurring multispecies cultures within existing ecologies show high resilience through symbiotic coexistence. Nevertheless, it remains a great challenge to realize such multispecies cultures in industrial applications. To address this challenge, multi-scale models, which integrate the metabolic information available from high-throughput experiments with the ecological scale of the species interaction and the process scale, have to be developed.

The goal of this study is to introduce a new formulation for dynamic flux balance models that explicitly accounts for obligatory interspecies interaction.

As a structured model for processes involving microorganisms, dynamic flux balance analysis (DFBA) combines the flux balance analysis models (Palsson, 2006) used in metabolic engineering with the dynamic mass balances of the reactor. Multispecies dynamic flux balance models have been previously studied in (Zhuang et al., 2011; Hanly and Henson, 2011). The model description assumes accumulation of the external metabolites determined by the production and uptake rates of the consortia members and does not explicitly consider dependency between the species.

The interaction between microorganisms can be direct communication, using signaling molecules or cell-cell contact, or indirect, through unidirectional or bidirectional exchange of external metabolites. The possible indirect interactions between each pair of organism are mutualism (each organism provides essential nutrients for the other organism), commensalism (only one organism depends on the metabolic product of the other), or neutralism (no direct dependency between the two organisms, but potential competition for the same resource).

In the context of DFBA, obligatory interactions, such as synthropy, can be realized by assuming that the metabolites that are being exchanged do not accumulate in the environment. This enforces additional constraints on the overall process model. The formulation can be interpreted from a game-theoretical perspective in which each player is represented by a microorganism with cost function given by the FBA model and the overall goal is to satisfy all constraints on the external metabolite concentrations given by their interconnection. A similar formulation for steady-state FBA models has been presented by Zomorrodi and Maranas (2012).

Numerical implementation is based on our previous work on simulation of dynamic flux balance models (Hoffner et al., 2013). Enforcing microbial interactions in dynamic models amounts to formulating a differential-algebraic system, in which the interaction constraints on the exchange fluxes are represented as algebraic equations. The linear programs representing the flux balance models of the members of the consortia are interconnected through the algebraic constraints and have to be solved simultaneously.

As a case study, consider a dynamic simulation of a thermophilic phototrophic mat system in Yellowstone National Park (Taffs et al., 2009). The model consist of oxygenic phototrophs, filamentous anoxygenic phototrophs, and sulfate-reducing bacteria represented by three guilds, which each represent a group of species that exploit the same class of environmental resources in a similar manner. There are several synthrophic interconnections between different members of the consortia, which are also changing from day time to night time. A 24h simulation of the mat system is implemented by sequentially simulating the two independent models accounting for diel (day-night)

References

T. J. Hanly and M. A. Henson. Dynamic flux balance modeling of microbial co-cultures for efficient batch fermentation of glucose and xylose mixtures. Biotechnology and Bioengineering, 108(2):376-385, 2011.

K. Hoffner, S. M. Harwood, and P. I. Barton. A reliable simulator for dynamic flux balance analysis. Biotechnology and Bioengineering, 110(3):792-802, 2013.

B. O. Palsson. Systems Biology: Properties of Reconstructed Networks. Cambridge University Press, New York, NY, 2006.

R. Taffs, J. Aston, K. Brileya, Z. Jay, C. Klatt, S. McGlynn, N. Mallette, S. Montross, R. Gerlach, W. Inskeep, D. Ward, and R. Carlson. In silico approaches to study mass and energy flows in microbial consortia: a syntrophic case study. BMC Systems Biology, 3(1):114, 2009.

K. Zhuang, M. Izallalen, P. Mouser, H. Richter, C. Risso, R. Mahadevan, and D.R. Lovley. Genome-scale dynamic modeling of the competition between rhodoferax and geobacter in anoxic subsurface environments. The ISME Journal, 5:305316, 2011.

A.R. Zomorrodi and C.D. Maranas. Optcom: A multi-level optimization framework for the metabolic modeling and analysis of microbial communities. PLoS Computational Biology, 8(2):e1002363, 2012.

(190e) Physiological, Molecular, and in Silico Analysis of Lipid Accumulation in the Marine Diatom Phaeodactylum Tricornutum

Mus, F.,
Carlson, R. P., Montana State University
Originally presented on: 11/4/2013 16:27:00 - 16:45:00

Photosynthesis derived lipids are considered promising precursors for sustainable biodiesel production. A detailed physiological, molecular, and *in silico* analysis of lipid accumulation under a suite of conditions including nitrogen limitation, alkaline pH stress, bicarbonate supplementation and organic acid supplementation was performed on the marine diatom *Phaeodactylum tricornutum*. For all tested conditions, nitrogen limitation was a prerequisite for lipid accumulation and the other culturing strategies only enhanced accumulation highlighting the importance of compounded stresses on lipid metabolism. Volumetric lipid levels varied depending on condition; the observed rankings from highest to lowest were for inorganic carbon addition (15 mM bicarbonate), organic acid addition (15 carbon mM acetate) and alkaline pH stress (pH 9.0). The results were incorporated into a genome-enabled stoichiometric model of the central metabolism of this diatom. Analysis suggested the central metabolism pathways associated with bicarbonate transport, carbonic anhydrases and C4 carbon fixation were important for lipid accumulation. Transcriptomic data also suggested that repurposing of phospholipids may play a role in lipid accumulation. This study provides a detailed physiological, molecular-level, and *in silico* foundation for improved understanding of diatom nutrient cycling and contributes to a metabolic blueprint for controlling lipid accumulation in diatoms.

(190f) Towards a Dynamic Understanding of Pathogen Detection and Response

Shoemaker, J. E., Japan Science and Technology Agency
Fukuyama, S., Japan Science and Technology Agency
Kawaoka, Y., Japan Science and Technology Agency
Kitano, H., Systems Biology Institute
Originally presented on: 11/4/2013 16:45:00 - 17:03:00

Towards a Dynamic Understanding of Pathogen Detection and Response

Pathogen detection and the ensuing immune response play a direct role in determining the severity of an infection. Recently, our group has identified key signaling mechanisms that determine when and to what degree the innate immune response is activated during infection with type A influenza viruses and showed that the degree of the immune response was associated with the tissue pathology. This allowed us to create the first mathematical model that can predict the timing and magnitude of the innate immune response to influenza infection which we validated using *de novo* infection data. We now report on our efforts to extend this model to be able to predict the host response to additional types of virus. We used specially constructed mutant influenza viruses whose ability to antagonize the immune response (by limiting interferon production) has been reduced. These viruses result in an early immune response relative to their respective wild-type viruses and tend to cause a less severe disease. From mice infected with the mutant viruses, we collected gene expression and protein data and determined the quantity of virus. We have found that only a slight change of a single parameter in the original model was necessary to predict the dynamics of the host response in the new infections. The change can be linked to the different binding affinities between the viruses and a key host protein – thus the model can be re-parameterized with a minimal number of experiments. The ability to develop an accurate model of the host response is the first step in developing advanced therapies which seek to manipulate the immune response – as opposed to directly targeting the virus – to manage disease severity and we believe such therapies will be more robust to virus mutations and applicable to a wide range of pathogens.

(190g) From Viral Sequences to Fitness Landscapes: A New Paradigm for in Silico Vaccine Design

Ferguson, A. L., University of Illinois
Hart, G. R., University of Illinois
Barton, J. P., MIT
Mann, J. K., University of Kwazulu-Natal
Omarjee, S., University of Kwazulu-Natal
Ndung'u, T., University of Kwazulu-Natal
Walker, B. D., Ragon Institute of Massachusetts General Hospital, Massachusetts Institute of Technology, and Harvard
Chakraborty, A. K., Massachusetts Institute of Technology
Originally presented on: 11/4/2013 17:03:00 - 17:21:00

Of the 34 million people living with HIV/AIDS, 23 million reside in the poor nations of sub-Saharan Africa that are least able to counter this epidemic. Hepatitis C virus (HCV) afflicts around 150 million people worldwide, and is the leading cause of liver transplantation in the developed world. No effective HIV or HCV vaccines exist. An inexpensive prophylactic vaccine offers the best hope to curb these epidemics, but systematic means to guide the design of effective vaccine immunogens for these, and other, infectious diseases are unavailable.

Appealing to spin glass models in statistical physics, we present a novel approach to translate viral sequence databases into landscapes of viral fitness described by Ising and Potts Hamiltonians. These landscapes chart the peaks and valleys of viral fitness as a function of amino acid sequence, and can be used to rationally design vaccines to eject the virus from the high fitness peaks, and drive it into the valleys where its compromised fitness impairs its ability to replicate and inflict damage to the host.

We illustrate this approach in the development of fitness landscapes for particular HIV and HCV proteins known to be particularly susceptible to immune pressure and drug therapies. In comparisons to experimental and clinical data, our inferred landscapes demonstrate good agreement with: 1) in vitro replicative fitness measurements, 2) clinically observed highfitness circulating viral strains, 3) documented HLA associated CTL escape mutations, and 4) intrahost viral mutation pathways revealed by deep sequencing. We demonstrate the value of data driven fitness landscapes in the *in silico* design of cytotoxic T-cell vaccine immunogens against HIV-1 clade B. We are currently working with experimental and clinical collaborators to test our designed immunogen in mice models of HIV infection.

(190h) Linking Phenotypes to Genotypes By Population-Based Mathematical Algorithms and Clustering Methods

San Miguel, A., Georgia Institute of Technology
Shen, K., Stanford University
Lu, H., Georgia Institute of Technology
Originally presented on: 11/4/2013 17:21:00 - 17:39:00

Since the advent of fluorescent biomolecules to study gene expression as well as protein localization and function, biological studies have relied heavily on microscopy. Furthermore, current high-throughput techniques enable the acquisition of large data sets where the relevant information is hidden and difficult to analyze. One approach to tackle these large data sets is to rely on mathematical and statistical approaches to extract relevant information, particularly in problems where phenotypes are subtle and the characteristic features are not known *a priori.*

Here, we take advantage of logistic regression models and clustering algorithms to analyze large data sets of phenotypic profiles of synaptic patterning in *C. elegans.* Recently, we developed high-throughput technologies that enable quantitative characterization of subcellular fluorescent reporters of synapses in live *C. elegans.* By incorporating microfluidics, computer vision and automation, we performed automated image acquisition and genetic screens where animals with interesting phenotypes are isolated. Particularly, we applied this system to find mutants with very subtle phenotypic differences in synaptic patterning. The phenotypic differences of the novel mutants range from a reduction in the number of micron-sized synaptic puncta, to subtle differences in puncta density. However, without prior knowledge of the phenotypic differences, establishing whether these are true mutants and where these differences may reside is exceedingly difficult when dealing with a multidimensional feature framework. Moreover, due to the stochastic nature of gene expression and environmental noise, isogenic populations of animals show a spread in the phenotypic landscape, making the identification of true subtle mutants exceedingly difficult. This is especially true when the features of interest are micron-sized synaptic sites. From our automated genetic screens we have obtained several putative mutants. With the aim of elucidating whether these are true mutants of interest, and predicting which genetic pathways could be altered in them, we apply two separate mathematical approaches to analyze phenotype-to-genotype relationships from a whole-population phenotypic profile perspective. Through statistical and mathematical analysis we are able to discern whether population-based differences are significant and can thus discriminate true subtle mutant phenotypes from false positives. Moreover, we apply custom clustering techniques to identify whether the true mutants identified correlate to known genetic pathways.

This method enables identification of genotypes of interest by comparing whole-population phenotypic profiles, while allowing prediction of putative altered genetic pathways.

(276a) Genome Engineering of Escherichia Coli Enhances Site-Specific Incorporation of Unnatural Amino Acids in Cell-Free Protein Synthesis

Hong, S. H., Northwestern University
Kwon, Y. C., Northwestern University
Martin, R. W., Northwestern University
Marcum, R.,
Des Soye, B. J.,
Jewett, M. C., Northwestern University
Originally presented on: 11/5/2013 8:30:00 - 8:48:00

Site-specific incorporation of unnatural amino acids (uAAs) into proteins and biopolymers by amber suppression makes possible new chemical properties, new structures, and new functions. However, competition between release-factor 1 (RF1) and orthogonal tRNAs have limited the technology. Here, we describe the development of a high yielding cell-free protein synthesis (CFPS) platform from genomically recoded *Escherichia coli* strains lacking RF1. Specifically, occurrences of the amber stop codon TAG are reassigned to the synonymous TAA codon. Because the parent strain has not been previously optimized for CFPS, we exploited multiplex automated genome engineering (MAGE) to design and construct synthetic genomes that upon cell lysis lead to improved extract performance. To stabilize template DNA, messenger RNA, amino acids, and protein products in CFPS, we targeted the deletion of 17 potential negative effectors. More than 50 strains were generated and tested in CFPS, allowing us to catalogue the systems impact of making numerous gene deletions both individually and in combinations. The protein synthesis activities of our most productive cell extracts were more than five-fold greater as compared to the extract from the parent strain, achieving more than 1.2 mg/ml of superfolder green fluorescent protein (sfGFP). The yields of modified sfGFP containing *p*-acetylphenylalanine at single and multiple positions in our *in vitro* system were more than ten-fold higher than the reported *in vivo* results. Our work demonstrates improved CFPS activity using whole genome editing for extract strain development and promises CFPS systems as suitable platforms for generating unnatural biopolymers.

(276b) Impact of Recovery Time On Recombination Efficiency in Escherichia Coli

Reynolds, T. S., University of Colorado
Gill, R. T., University of Colorado
Originally presented on: 11/5/2013 8:48:00 - 9:06:00

With the rapid advancement of DNA synthesis technologies, the ability to efficiently recombineer user-defined DNA into organisms of interest has become increasingly important. Recombination efficiency in *Escherichia coli* has been the subject of many studies. Despite these efforts, there is still considerable variability in reported recombination efficiencies. We hypothesized that multiple chromosome copies within *E. coli* could partially explain reported variability. With more than one chromosomal target per *E. coli* cell, population phenotype expression levels could change as recombineered chromosomes are segregated to daughter cells. A simple predictive tool was developed to estimate cell population phenotype and genotype during outgrowth after recombination. We found that recombineered chromosomes were increasingly segregated within the population; leading to a dilution of the phenotype of interest but an increase in chromosomal homogeneity. These results were confirmed experimentally leading to the conclusion that recovery time should be accounted for when reporting recombineering efficiencies.

(276c) White-Boxing Genetic Circuit Modeling: Absolute TF Binding Free Energies From Fluorescence Measurements

Farasat, I., Pennsylvania State University
Salis, H., Pennsylvania State University
Originally presented on: 11/5/2013 9:06:00 - 9:24:00

Every genetic circuit model in existence today has been parameterized with proportional data. Fluorescence. Absorbance. Luminescence. These measurements are only proxies for the cell's internal protein concentrations, and yet we use them to record the cell's internal dynamics and to measure the biophysics of its genetic parts. A first-principles biophysical model that uses proportional data will always be left with an unknown proportionality constant, and this unknown factor will depend on many extrinsic and ever-changing variables. Beyond the creation of community-wide measurement standards, we investigated whether a set of properly designed experiments could measure the *instrinsic* biophysical properties of a genetic part inside the cell, using only proportional fluorescence as the measurement output. We then asked whether this intrinsic knowledge could be used to predict the behavior of a family of synthetic genetic circuits.

Specifically, we develop a high-throughput quantitative approach to measuring the absolute binding free energies of a transcription factor to its DNA operator sites, utilizing only protein reporters and fluorescence measurements. The expression output of the transcription factor's regulated promoter is recorded within a series of rationally designed genetic sandboxes that contain additional non-regulating TF-binding sites. Our first-principles quantitative model shows, and we experimentally demonstrate, that the ratio in fluorescence measurements between two genetic sandboxes yields a direct, absolute relationship to the TF's binding free energy, as measured in kcal/mol.

We validate this approach using three newly engineered repressors, measuring their TF binding free energies, and predicting the behaviors of entire families of genetic circuits. We validate these predictions experimentally and demonstrate that the resulting biophysical model of a TF-regulated promoter can predict its transcription rates across a wide range of TF expression levels. We use these models to identify optimal NOT-gate genetic circuits with maximal dynamic ranges and tunable input thresholds. The affirmative answers to our questions take us two steps closer to creating a genetic circuit engineering discipline that is free of context effects.

(276d) Toward Practical Applications of Synthetic Biology: Part Mining to Construction of Programmable Cells

Shopera, T.,
Immethun, C.,
Henson, W. R.,
Moon, T. S., University of California, San Francisco
Originally presented on: 11/5/2013 9:24:00 - 9:42:00

The past decade has witnessed the power of synthetic biology in the creation of synthetic devices and systems that help to understand complex biological systems. However, its potential for real-world applications has not been fully exploited. One of its promising applications is the construction of programmable cells that are able to integrate multiple environmental signals and to implement synthetic control over biological processes. Here, we harness a toolbox of genetic parts from bacteria to construct genetic circuits. To this end, genetic parts have been recruited from various organisms and assembled into genetic devices, which were characterized using fluorescent proteins as outputs. Our programmable cells contain environmental sensors to distinguish various conditions and multi-input logic programs with genetic memory to generate the output only when the user-defined conditions are met. We will present progress on the construction of such engineered cells toward practical applications, such as disease prevention and chemical production.

(276e) Rational Design of Synthetic Yeast Promoters Using Systems-Level Bioinformatics Models

Curran, K., The University of Texas at Austin
Crook, N., The University of Texas at Austin
Karim, A., The University of Texas at Austin
Gupta, A., The University of Texas at Austin
Alper, H., The University of Texas at Austin
Originally presented on: 11/5/2013 9:42:00 - 10:00:00

Promoters are a key component in metabolic engineering and synthetic biology. These elements enable expression of heterologous genes, provide regulation of genetic circuits, and establish the basis for wholly rewired cells capable of producing new biochemical products. As such, a large amount of effort has been expended to characterize and synthesize both endogenous and synthetic promoters. Traditionally, new promoters have been generated through random mutagenesis of native promoters, bioprospecting to find non-native promoters, and synthetic hybrid promoters. However, with the exception of the latter example, little progress has been made to rationally and systematically engineer promoters for a prescribed activity level. In this presentation, we describe a new approach—namely a model-driven promoter design tool. Specifically, we highlight recent progress in using models to rationally re-design endogenous yeast promoters. In each of these cases, the rationally designed promoters followed the expected transcriptional output. This work therefore represents a fundamentally new method for understanding promoter architecture with regards to expression level. Moreover, this work enables rational design of synthetic parts.

(276f) Functional Characterization of Saccharomyces Cerevisiae Promoters and Application in Pathway Engineering

Li, Z.,
Zhang, G.,
Li, C., Beijing Institute of Technology
Originally presented on: 11/5/2013 10:00:00 - 10:18:00

Synthetic biology is considered as an emerging research field that will bring new opportunities to biotechnology. Among the microbial platforms in synthetic biology, *Saccharomyces cerevisiae* is one of the most widely used organisms due to its well-characterized physiology and genetics, fast cell growth rate, and availability of abundant genetic tools.

Most metabolic engineering and synthetic biology studies have focused on selecting the appropriate enzymes, because enzyme homologs cloned from a variety of organisms could have different behavior in terms of expression and activity in a heterologous host. However, selection of proper promoters is also important especially for yield and productivity optimization. This is not only because the protein transcription level is determined by the promoter strength, but also because promoters could be regulated and behave differently under different growth conditions. The constitutive promoters have been reported mostly come from housekeeping genes, including glycolysis pathway genes and elongation factor. Here we report the cloning of eighteen constitutive promoters employing eGFP as a reporter from the *S. cerevisiae* genome and the comparison of their strengths. Five of them come from glycolysis pathway genes, the rest thirteen come from amino-acyl tRNA synthetase genes. To our knowledge, this is the first time report about the promoters that come from amino-acyl tRNA synthetase genes. By analyzing the sequence of the promoters, we found some transcription factors could affect the strengths of promoters. In the batch fermentation, the strengths of promoters that come from different pathways behave in different patterns. We chose the promoters with relatively high strength and cloned the xylan degradation pathway and xylose reductase genes under these promoters, and transformed into *S. cerevisiae*. The resulting strain synthesized xylitol directly from beechwood xylan, and therefore can be utilized as a consolidated bioprocessing (CBP) organism for biomass utilization and conversion. During the fermentation we found the promoter not only affected the expression of gene, but also the growth of the cell.

(276g) Sensing- and Communication-Dependent Control of Microbial Motility

Bower, A. G. W., Rensselaer Polytechnic Institute
Julius, A. A.,
Collins, C. H., Rensselaer Polytechnic Institute
Originally presented on: 11/5/2013 10:18:00 - 10:36:00

The long-term goal of this project is to enable the use of microorganisms for the controlled motion of micron-scale devices, known as microrobots. A key aspect of his design is the generation of strain of *E. coli* where motility is tightly regulated by a cell-cell communication, or quorum sensing (QS) signal. The second biological module will be a strain, or combination of strains, that recognize extracellular signal(s) and produce the QS signal recognized by the first strain. This approach is advantageous because the sensor strains can easily by swapped to recognize different signals or combined to recognize multiple signals, or perform more complex computation. The use of QS may also be useful for amplifying the signal when low concentrations of a target molecule are present. We have constructed a strain of *E. coli*, where its swimming motility is tightly controlled by the *esa* QS system. The esa system provides advantages over the well-known *lux* system because its regulator, EsaR, can act as a repressor, which leads to lower levels of background expression. This is expecially important when controlling genes where only a few copies can enable a change in phenotype. In this case, we have used EsaR to control the expression of a key motor protein. The ribosome binding site was also engineered to optimize the control of the expression of the motor protein and enable QS-dependent motility. In swim plate assays, we have shown that our new construct shows no swimming motility in the absence of QS signal. Interestingly, we have also observed directional swimming, where the swimming cells move up a gradient of QS signal generated by a second strain of cells producing the signal.

(276h) Highly Orthogonal Programmable Riboregulators for Complex Digital Logic in E. Coli

Green, A. A., Harvard University
Silver, P., Harvard Medical School
Collins, J. J., Howard Hughes Medical Institute, Boston University
Yin, P., Harvard Medical School
Originally presented on: 11/5/2013 10:36:00 - 10:54:00

A diverse array of synthetic genetic programs has been constructed inside living cells to carry out digital logic operations[1], record memories of biochemical events[2], and synthesize valuable chemicals[3,4]. These behaviors have been established largely using existing biological components rewired in novel ways to enable new capabilities. It remains a significant challenge, however, to engineer living systems with functionalities that extend beyond rudimentary logic operations. This challenge arises in part from the limited number of biological circuit components that operate independently from one another and the difficulty in creating robust approaches to combine these elements for predictable behavior in living cells. Here, we exploit synthetic networks of RNA self-assembly reactions to construct a versatile platform for complex digital logic in *Escherichia coli*. The building blocks of this computational platform are newly designed programmable riboregulators of gene expression consisting of pairs of cis-repressed mRNAs (crRNAs) and trans-activating RNAs (taRNAs), which hybridize with their cognate crRNAs to de-repress them. In contrast to previous work, these RNA-based regulatory systems exploit design concepts derived not from existing biological systems[5-7] but from hairpin-based strand displacement reactions developed in nucleic acid nanotechnology[8,9]. These design features greatly expand the RNA sequence space of potential riboregulators compared to earlier systems[7,10] and provide the riboregulators with a high degree of programmability, orthogonality, and modularity for carrying out digital logic. To demonstrate these capabilities, we have assembled a family of 48 programmable riboregulators that increase expression of a GFP reporter by a factor of 40 or more upon transactivation and have validated a set of 15 highly orthogonal riboregulators with cross-talk levels below 10% *in vivo*. We have used this expansive toolkit of synthetic biological components to construct a number of complex decision-making systems in living cells. These include a multi-input OR gate that activates protein translation only in response to one of six programmed taRNAs and an AND circuit that produces GFP only upon assembly of a two-strand RNA complex. These programmable riboregulators provide a robust platform for evaluating more complex computations in living cells, and also constitute a promising test bed for probing interactions between assemblies of multiple RNA molecules *in vivo*.

References

1. Moon, T. S., Lou, C., Tamsir, A., Stanton, B. C. & Voigt, C. A. Genetic programs constructed from layered logic gates in single cells. *Nature* **491**, 249-253 (2012).

2. Friedland, A. E., Lu, T. K., Wang, X., Shi, D., Church, G. & Collins, J. J., *Science* **324**, 1199-1202 (2009).

3. Steen, E. J., Kang, Y., Bokinsky, G., Hu, Z., Schirmer, A., McClure, A., del Cardayre, S. B. & Keasling, J. D., *Nature* **463**, 559-U182 (2010).

4. Delebecque, C. J., Lindner, A. B., Silver, P. A. & Aldaye, F. A., *Science* **333**, 470-474 (2011).

5. Isaacs, F. J., Dwyer, D. J., Ding, C. M., Pervouchine, D. D., Cantor, C. R. & Collins, J. J., *Nature Biotech.* **22**, 841-847 (2004).

6. Lucks, J. B., Qi, L., Mutalik, V. K., Wang, D. & Arkin, A. P., *Proc. Natl. Acad. Sci. USA* **108**, 8617-8622 (2011).

7. Mutalik, V. K., Qi, L., Guimaraes, J. C., Lucks, J. B. & Arkin, A. P., *Nature Chem. Biol.* **8**, 447-454 (2012).

8. Dirks, R. M. & Pierce, N. A., *Proc. Natl. Acad. Sci. USA* **101**, 15275-15278 (2004).

9. Yin, P., Choi, H. M. T., Calvert, C. R. & Pierce, N. A., *Nature* **451**, 318-322 (2008).

10. Callura, J. M., Cantor, C. R. & Collins, J. J., *Proc. Natl. Acad. Sci. USA* **109**, 5850-5855 (2012).

(333a) Synthetic Cell-Based Biosensors for Dynamically Imaging Immune Function in Vivo

Daringer, N., Northwestern University
Dudek, R. M., Northwestern University
Leonard, J. N., Northwestern University
Originally presented on: 11/5/2013 12:30:00 - 12:48:00

The ability to dynamically monitor or visualize local immune states *in vivo* would transform our ability to understand processes such as the establishment and development of dysfunctional immune states at tumor sites and the response of these multicellular networks to potential therapeutic interventions. To date, however, we are limited to systemic measures (such as profiles of serum cytokines or circulating immune cells) or terminal assays (requiring tissue biopsy or, more commonly, sacrifice of the experimental animal). We have established a novel mammalian synthetic biology technology enabling us to engineer cells into living "biosensors" that are responsive to environmental stimuli that are exclusively extracellular, such as the cytokines that characterize specific immunological states. These cell-based biosensors detect a relevant analyte of interest and then produce an optical signal that can be imaged in whole, living animals (e.g., luminescence, near-infrared fluorescence). This approach is fundamentally distinct from a reporter gene approach, in which one can only detect whether a given gene is expressed in the engineered cell, not whether the engineered cell is exposed to the analyte, and therefore this approach provides information not currently accessible. In work enabling this approach, we have developed a modular extracellular sensing architecture (MESA) that can be adapted to diverse applications and that facilitates optimization of performance characteristics.

We have previously shown that MESA biosensors signal in a dimerization-dependent fashion, and this mechanism was used to sense the cytokine IL-10, which plays an important role in tumor-related immune dysfunction. Here, we first describe design innovations that improve signal-to-noise ratio and robustness to facilitate translation to *in vivo* applications. As an initial proof of principle, we engineered tumor cells to function as biosensors, such that following adoptive transfer of engineered tumor cells to a syngeneic mouse host, the tumor monitors and reports upon its own immunological milieu. We are initially applying this technology to help understand how the immune response shifts from anti-tumor (immunostimulatory) to pro-tumor (immunosuppressive) during the course of tumor progression. MESA technology can be extended to generate cell-based therapies by coupling biosensing to the expression of therapeutically-relevant gene products or cellular functions including intracellular gene or protein circuits that perform logical evaluation of multiparametric inputs. These proof-of-principle experiments thus provide a quantitative framework for engineering cell-based devices that perform customized functions *in vivo* to enable both scientific inquiry and novel therapeutic strategies.

(333b) Synthetic Circuits Design Using Proteon and Proteoff

Biliouris, K., University of Minnesota
Kaznessis, Y., University of Minnesota
Originally presented on: 11/5/2013 12:48:00 - 13:06:00

Even though recent advances in synthetic biology have fueled our ability to build synthetic systems with user-defined function, the field still faces central challenges. One of the main challenges is the limited level of complexity underlying the newly designed synthetic systems; this complexity has recently reached a plateau [1]. Effectively addressing this limitation may necessitate the use of validated mathematical models that inform the design of synthetic systems with high complexity.

We have previously built and characterized two synthetic devices, proTeOn and proTeOff, that tightly control prokaryotic gene expression in a tetracycline-dependent fashion [2]. proTeOn (proTeOff) upregulates the expression of a target gene in response to the presence (absence) of tetracycline. We are now using these devices to build synthetic circuits with various potential dynamics ranging from oscillations to bistability and switch-like behavior. These circuits integrate proTeOn and proTeOff along with molecular components from the *lactose* and *tetracycline* operon. We are currently designing these systems *in silico* using our experimentally validated models [2,3,4]. Mathematical modeling will then guide the *in vivo*design and characterization of these systems.

These synthetic circuits will contribute towards designing systems with increased complexity which are robustly controlled and composed of well-studied molecular parts.

1) Purnick, P. E., & Weiss, R. (2009). The second wave of synthetic biology: from modules to systems. *Nature Reviews Molecular Cell Biology*, *10*(6), 410-422.

2) Volzing, K., Biliouris, K., & Kaznessis, Y. N. (2011). ProTeOn and ProTeOff, new protein devices that inducibly activate bacterial gene expression. *ACS Chemical Biology*, *6*(10), 1107-1116.

3) Biliouris, K., Daoutidis, P., & Kaznessis, Y. N. (2011). Stochastic simulations of the tetracycline operon. *BMC systems biology*, *5*(1), 9.

4) Ramalingam, K. I., Tomshine, J. R., Maynard, J. A., & Kaznessis, Y. N. (2009). Forward engineering of synthetic bio-logical AND gates. *Biochemical Engineering Journal*, *47*(1), 38-47.

(333c) Loads and Modularity in Synthetic and Natural Switches: A Computational Study

Lyons, S. M.,
Xu, W.,
Medford, J.,
Prasad, A., Colorado State University
Originally presented on: 11/5/2013 13:06:00 - 13:24:00

We are on the cusp of significant breakthroughs in the use of synthetic gene circuits for a variety of applications. However the mathematical design of simple synthetic circuits is often made under assumptions of modularity, in particular the idea that these modules will behave identically in isolation and in connection. Similarly analysis of signal transduction or gene transcription networks is also carried out under the assumption of modularity, allowing analysis of motifs in isolation from the rest of the network. A limitation of this modularity assumption is the observation that downstream protein circuits affect the properties of an upstream circuit, without explicit feedback, because of binding between the interconnected parts of the two modules (Sontag et. al. Mol. Sys. Biol., 4:161). Thus understanding the effects of adding additional elements to the output of these technologically important network modules is required for a thorough understanding of the challenges of scaling up synthetic networks to higher levels of complexity. In this study we examine the qualitative behavior of the synthetic genetic toggle switch, as well as a toggle switch motif that we identify in the mammalian apoptosis network and other natural systems, and finally positive feedback switch motifs that have been shown to play a significant role in T cell activation. We show that the addition of a simple binding partner, i.e. a load, to the output of these circuits has the potential to change the qualitative behavior of the circuits. For the simple genetic toggle switch we find that loads affect the dynamics of switching between the two states. We construct the quasi-potential landscape of the toggle switch and show that loads skew the potential landscape dramatically. We show that some naturally occurring toggle switches have an additional motif that in principle allows the system to tune the quasi-potential landscape to deal with differing loads. We also show that for some naturally occurring signal transduction switches based on positive feedback, loads can abrogate switch-like behavior completely. Our analysis underscores the importance of analyzing the effect of connecting elements when studying network modules.

(333d) A Mammalian Synthetic Gene Circuit to Quantify Activation of the Ubiquitin Proteasome System

Zhao, W., Rice University
Bonem, M.,
McWhite, C.,
Segatori, L., Rice University
Originally presented on: 11/5/2013 13:24:00 - 13:42:00

The ubiquitin proteasome system (UPS) is the main protein quality control system in eukaryotic cells that regulates degradation of damaged or misfolded proteins. Inefficient or defective UPS activity results in accumulation of misfolded proteins and aggregation, which are the hallmarks of a number of neurodegenerative diseases, including Parkinson's and Alzheimer's. Protein aggregates are also key impediments to the high expression of recombinant proteins and limit the high-yield production of proteins for research, diagnostic and industrial applications. We hypothesized that enhancing proteasomal degradation would facilitate intracellular clearance of misfolded proteins and prevent accumulation of protein aggregates. A number of proteasome inhibitors in the form of small molecules and peptides that bind to the catalytic sites of the proteasome have been reported. They have been widely used to characterize the role of the UPS in maintaining protein homeostasis and recently even transitioned to clinical therapeutics. On the other hand, pharmacologic agents that enhance UPS activity are rare and still largely unexplored. To fill this knowledge gap and generate a platform for high-throughput screening of proteasome activators, we developed a mammalian orthogonal genetic circuit that interfaces with cell's natural proteasomal degradation pathways. In this synthetic circuit, increase of UPS activity is coupled with readily detectable fluorescence signal. The architecture of the genetic circuit is based on a transcriptional repressor engineered to function as a UPS model substrate. Guided by predictive modeling, we refined the circuit architecture to achieve a more robust and decisive response to modulation of proteasomal degradation. We report experimental evidence that self-activation of the repressor creates a positive feedback loop that results in enhanced sensitivity of the circuit to UPS activity and larger output signal window. This novel strategy overcomes limitations of current methods, which are based on detecting a decrease in concentration of a degradation-prone signal substrate and are prone to artifactual results.

(333e) Selective Killing of Bacteria With the Crispr Immune System

Gomaa, A. M.,
Beisel, C. L., North Carolina State University

Originally presented on: 11/5/2013 13:42:00 - 14:00:00

One of the greatest challenges facing modern medicine is the rise and spread of multidrug resistance. Most research efforts have focused on developing novel classes of antibiotics to supplement our current arsenal. However, these same antibiotics will be subject to the same mechanisms of resistance, providing at best a temporary solution to this growing epidemic. A promising alternative is the development of novel treatment strategies that circumvent common mechanisms of resistance. Here, we describe such a strategy that, unlike existing antibiotics, can differentiate between pathogens and the multitude of beneficial bacteria within our bodies. Our strategy relies on "tricking" the CRISPR immune system into recognizing the genome as a foreign invader, inducing genome destruction and bacterial cell death. Recognition is conducted by CRISPR RNAs that can be designed to target almost any genomic sequence. Using existing databases of genomic sequences, CRISPR RNA sequences can be rationally designed that differentiate between harmful and beneficial bacteria. Using the Type I-E CRISPR system in *Escherichia coli* as a model, we observed potent and selective killing of *E. coli* and *Salmonella typhimurium*. We further found that CRISPR RNAs could be designed to readily differentiate between even between closely related strains. These findings offer alternative strategies in the ongoing fight against multidrug resistance. Furthermore, our findings may lead to novel tools in systems biology for the analysis of complex microbial consortia, the evolution of interactions between bacteria and bacteriophages, and the system-wide response to double-stranded DNA breaks.

(333f) Detecting and Killing Pathogenic Bacteria With Lactic Acid Bacteria

Volzing, K. G., University of Minnesota
Borrero, J., University of Minnesota
Dunny, G. M., University of Minnesota
Kaznessis, Y. N., University of Minnesota
Originally presented on: 11/5/2013 14:00:00 - 14:18:00

Over the last several decades *Enterococcus* has undergone a troublesome transformation from being a harmless, if not helpful, commensal organism to a major public health threat. Vancomycin-resistant *Enterococcus* (VRE) are now the primary cause of hospital-acquired infections and no longer susceptible to our most potent antibiotics. This problematic emergence of VRE has motivated us to develop a unique cell-based seek-and-destroy system that first detects *Enterococci* and then targets them for cell lysis and growth inhibition. To do this, we have cloned the regulatory components of the prgX/prgQ operon from *Enterococcus* into *Lactococcus lactis*. These promoters are controlled by a cell-cell communication system where transcription is upregulated by cCF10. This peptide pheromone is constitutively produced and secreted by most, if not all, *Enterococci* and functions as the signal for our system. Upon detection of cCF10, the promoter overexpresses the downstream antimicrobial peptide (AMP) genes that we have selected for *Enterococcus* lysis and growth arrest. The AMP products Nisin A, Enterocin A and Enterocin P are then secreted by *L. lactis* to target and kill *Enterococcus* by membrane disruption.

(333g) Model-Based Optimal Design of a Gene Metabolator System

Bhatti, A., University College London
Dua, V., University College London
Originally presented on: 11/5/2013 14:18:00 - 14:36:00

Model-based optimal design of a gene oscillator system

Asif Bhatti, Vivek Dua

Synthetic biology has garnered much interest in recent years and looks to revolutionise a wide range of industries including genetics and protein engineering. It is a novel, multidisciplinary approach utilising engineering principles and biology to enhance naturally occurring systems. First described by Szybalski and Skalka [1], it heralds a new age where genes can not only be described, but re-evaluated to give better arrangements of the gene network. Modelling and designing engineered biological systems to predict system performance before fabrication is an important component of synthetic biology [2,3]. In this sense it is similar to systems biology as they both rely heavily on computer modelling.

The main aim of this work is to optimally design a metabolator, which is a metabolic enzyme process of Acetyl-CoA and its co-factors. The metabolic process can be described using a dynamic model involving nonlinear differential equations [4].The metabolator consists of two metabolite pools, one with the enzyme and another with its co-factors, acetyl phosphate (AcP) and acetate kinase (Ack). The two pools are connected by a negative feedback reaction from the enzyme to the co-factors involving phosphate acetyltransferase (Pta), and a positive feed-forward reaction from the co-factors back to the enzyme involving acetyl synthase (Acs). The influx of glycerol, fatty acids and sugars (glycolytic flux) into the enzyme, Acetyl-CoA, causes its concentration to oscillate. As a result Pta feeds into the co-factor metabolite pool to produce acetate. Acetate is then transported to the cell for further intrinsic use. However the process is cyclical and can have a positive feed?forward reaction whereby the presence of acetate in the metabolite pool (AcP and Ack) can upregulate Acs formation to feed back to the enzyme. The process reacts to circadian rhythm in the body, and will continue to oscillate as a result.

The dynamic model of the metabolator is transformed into a system of simultaneous algebraic equations using an artificial neural network (ANN) transformation [5]. The metabolator design is formulated as an optimisation problem, where the objective is to control the concentration of Acetyl-CoA at a desired set-point by manipulating the glycolytic flux. Also, when acetate levels are increased to large values the system can be inhibited and will no longer oscillate, a property that could be useful for systems requiring a ?shut-off' feature. To ensure robustness of the numerical convergence of the optimisation problem, the solution of the dynamic equations was also verified by the fourth order Runge-Kutta method, and the effect of the numerical integration step-size was also analysed

The metabolator performance is subject to uncertain parameters such as the surrounding environmental conditions (pH, temperature, availability of nutrients and age) and degradation of the enzyme. In the future work, the effect of these uncertain parameters on the design will be analysed, and a new design that can effectively take into account the uncertain parameters and yet deliver the performance required will be obtained.

[1] Szybalski, W. & Skalka, A., *Nobel-Prizes and Restriction Enzymes*. Gene, 4, 181-182, 1978

[2] Dasika, M. S. & Maranas, C. D., *OptCircuit: An optimization based method for computational design of genetic circuits*. Bmc Systems Biology, 2, 2008

[3] Purcell, O., Savery, N. J., Grierson, C. S. & Di Bernado, M., *A comparative analysis of synthetic genetic oscillators*, Journal of the Royal Society Interface, 2010

[4] Fung, E., Wong, W. W., Suen, J. K., Butler, T., Lee, S. G. & Liao, J. C., *A synthetic gene-metabolic oscillator*. Nature, 435, 118-122, 2005

[5] Dua, V. & Dua, P., *A simultaneous approach for parameter estimation of a system of ordinary differential equations, using artificial neural network approximation*. Industrial and Engineering Chemistry Research, 51 (4), 1809-1814, 2012

(333h) An in-Silico Synthetic Approach to Elucidate Design Principles Underlying Bi-Directional Gradient Sensing in Eukaryotic Cells

Krishnan, J., Imperial College London
Alam-Nazki, A., Imperial College London
Originally presented on: 11/5/2013 14:36:00 - 14:54:00

Many cells (both bacteria and eukaryotes) respond to gradients of concentrations of chemicals in their environment by exhibiting directional migration, a process known as chemotaxis. This process plays very important biological and physiological roles. In many cases, the movement is up-gradient, and the process is referred to as chemoattraction, and in some other cases, the movement is down-gradient and the process is referred to as chemoorepulsion. It has been experimentally found that different eukaryotic cells are capable of both chemoattraction and chemorepulsion, either to different chemicals, or in some cases to the same chemical but under different conditions. It is thus important to examine what features in the wiring of the signalling networks and their organization allow the cell to exhibit both these features, and how this affects the signal transduction in each case. In this talk we will use an in-silico synthetic approach to examine different design features in the wiring of signalling networks to give rise to such behaviour, in light of known complexities in signalling in eukarytotic chemotaxis.

We do this by systematically analyzing two cases, motivated by their postulation and discussion in the experimental literature. The first is that of a “polarity switch” and the second is the role of competing signalling effects at the gradient sensing. For each of these cases, we will discuss how these basic network wiring possibilities may allow for the cell to exhibit attractive and repulsive response, and then examine how this works upstream of different qualitative signalling behaviour such as adaptation and spontaneous polarization. This is achieved by “connecting” the relevant modules in-silico. This systematic approach reveals the behaviour of each design feature, the capabilities and constraints involved. We will then discuss the implication of our various results in the context of the known signalling of different eukaryotic cell types. Overall this synthetic approach provides a design framework and "systems skeleton"for examining and elucidating design principles of attractive/repulsive gradient sensing in multiple systems.

(378a) Towards A Genetic Basis for Antibiotic Resistance in Chromobacterium Violaceum

Das, D.,
Raghunathan, A., National Chemical Laboratory
Originally presented on: 11/5/2013 15:15:00 - 15:33:00

The emergence of mutations in the genome is one of the major factors underlying evolution. Mutation as the cause of antibiotic resistance was studied in *Chromobacterium violaceum* for the antibiotics chloramphenicol and streptomycin. The wild-type *C. violaceum* was tested to be susceptible to low concentrations of both antibiotics (Chloramphenicol: MIC 8ug) (Streptomycin: MIC 10ug). Continuous exposure to the antibiotics at concentrations slightly lower than the MIC, resulted in resistance to the antibiotics. This adaptive evolutionary phenomenon allowed growth on agar plates at high antibiotic concentrations (MIC: 200ug). The growth rates of these evolved strains 2-30 fold higher at varying concentrations of antibiotic. All genetic changes were delineated during the evolution of antibiotic resistance using next generation sequencing (Ion Torrent^{TM}). This talk will focus on the results of NGS and the functional impact of the genetic variations detected including AcrAB drug efflux pumps and ribosomal proteins like rpsL.

(378b) The Small Rnome of Clostridium Acetobutylicum That Responds to Butanol and Butyrate Stress

Venkataramanan, K. P., University of Delaware
Jones, S., University of Delaware
McCormick, K. P., University of Delaware
Kunjeti, S. G., University of Delaware
Ralston, M., University of Delaware
Meyers, B. C., University of Delaware
Papoutsakis, E. T., University of Delaware
Originally presented on: 11/5/2013 15:33:00 - 15:51:00

In the genus *Clostridium*, which is of major importance to human and animal pathogenesis, human physiology, the carbon cycle, biorefinery and biotechnological application, very few regulatory non-coding small RNA (sRNA) have been so far identified. Significantly, no sRNAs have been reported as being involved in toxic metabolite stress or any stress. In *C. acetobutylicum*, we have previously predicted 113 sRNAs and have experimentally validated a subset of 31 of them. Here, using deep RNA sequencing (RNA-seq), we investigated the expression of the small RNome of *C. acetobutylicum* in response to two important toxic metabolites, butanol and butyrate. Butanol is the major metabolite of the solventogenic stationary phase while butyrate is the major product of the acidogenic growth phase. The goal of this study is to identify sRNAs that respond to butanol and/or butyrate stress, aiming to identify sRNAs that can be engaged to engineer strains tolerant to these toxic metabolites, but significantly also discover if and how sRNAs respond to metabolic stresses.

We collected a large set of temporal RNA-seq data from multiple experiments with biological replicates employing various levels of metabolite stress. We mined the RNA-seq data for robustly identifying the presence and differential expression of sRNAs by relying on the analysis, based on the RNA-seq data, of the previously predicted 113 and the validated subset of 31 sRNAS. This analysis enabled us to set metrics (read counts) for reliably identifying the presence of new sRNAs from RNA-seq data. Based on these metrics, we identified an additional set of 46 sRNAs not previously known or predicted for this organism. We then examined the differential expression under metabolite stress of all identified sRNAs and verified the presence of a select subset by Northern analysis. Based on the conservation of the putative Hfq binding module, we identified the subset of these sRNAs that are likely to be regulated by the well-conserved RNA chaperone Hfq. Finally, we examined the likely role of a few of these stress-related sRNAs by relating their expression pattern to identifiable physiological characteristics.

This is the first comprehensive study of the stress-related sRNome in a *Clostridium* organism. Elucidation of the role of sRNA in clostridial response to metabolite stress will be an essential component for understanding the complexity of the regulatory network that underlies these stress responses.

(378c) Discovery and Regulation of Biomass-Degrading Enzymes From Anaerobic Gut Fungi Using Next-Generation Sequencing

Solomon, K., University of California, Santa Barbara
Haitjema, C., University of California, Santa Barbara
Borges-Rivera, D., Broad Institute of Massachusetts Institute of Technology and Harvard
Thompson, D. A., Broad Institute of Massachusetts Institute of Technology and Harvard
Regev, A., Broad Institute of Massachusetts Institute of Technology and Harvard
O'Malley, M. A., University of California, Santa Barbara
Originally presented on: 11/5/2013 15:51:00 - 16:09:00

Lignocellulosic biomass is a renewable resource capable of supplying one third of the domestic liquid fuel demand. However, processes to extract fermentable sugars from this biomass involve expensive pretreatment and enzymatic digestion steps that rely on a limited number of characterized cellulases. Therefore, there is a critical need to identify new technologies for the economic conversion of crude biomass into sugars. Anaerobic fungi within the digestive tracts of large herbivores are powerful biomass degraders yet they remain a rich untapped source for new lignocellulolytic enzymes. Despite their lignocellulolytic behavior, the enzymes responsible are largely unknown due to limitations associated with the cultivation and molecular characterization of gut fungi. In this study, we sought to identify these enzymes through RNAseq, and capture their expression profiles to understand the hydrolytic behavior of gut fungi against lignocellulosic substrates.

We have isolated a novel gut fungus from the *Piromyces* genus, which we have employed as a model organism to study biomass degradation. The isolate exhibits strong cellulolytic activity against diverse cellulosic and lignocellulosic substrates, which is repressed in the presence of simple sugars (e.g. glucose). Using next generation strand specific RNAseq and the TRINITY bioinformatic platform, we have assembled the first reported *de novo* gut fungal transcriptome. Analysis of the transcriptome reveals transcripts encoding over a hundred novel lignocelluloytic enzymes with diverse catalytic activities, which are either freely secreted or associated with fungal cellulosome complexes. Furthermore, we have confirmed the biological presence of enzymes encoded by assembled transcripts, supporting the quality of the *de novo* assembly. Also embedded within the transcriptome are quantitative transcript abundance information and non-coding RNA such as antisense RNA, which we hypothesize to directly regulate the expression of these cellulases. In this talk, I will describe the subset of novel enzymes that we have identified, and discuss the global regulatory patterns observed under various growth conditions.

(378d) Microdevices and Materials for Genomic Analysis of Single Microbial Cells

Meagher, R. J., Sandia National Laboratories
Light, Y., Sandia National Laboratories
Yilmaz, S., Sandia National Laboratories
Rhee, M., University of Michigan
Singh, A. K., Sandia National Laboratories
Liu, P., Sandia National Laboratories
Originally presented on: 11/5/2013 16:09:00 - 16:27:00

Next-generation sequencing technologies have made whole-genome sequencing of microbes routine, as long as the microbe can be grown in pure culture to provide sufficient quantities of clonal DNA for library preparation. For microbes that are not cultured, which includes the vast majority of taxa both in the environment and within the human body, other techniques must be used to analyze microbial genomes. Metagenomics is useful for analyzing genes and pathways at the level of microbial communities, but assembly of complete microbial genomes from a mixed population is difficult except for the simplest communities. We have developed a suite of technologies that complement metagenomic techniques by analyzing the genomic content of single cells. We begin with complex microbial communities, isolate individual cells, and then amplify to prepare sequencing libraries. Key aspects of the technology include: (1) compartmentalization of single cells in nanoliter-volume droplets, as ideal "nanoreactors" for amplifying the genome of a single microbial cell, (2) multiple displacement amplification (MDA) using Phi29 DNA polymerase, with special consideration to purity requirements when amplifying single cells, and (3) labeling and sorting techniques based on fluorescence in situ hybridization (FISH), either before or after amplification, to identify interesting or novel microbes based on phylogeny (i.e. 16S rRNA sequence) or presence of functional genes. Our pipeline utilizes a combination of conventional instrumentation (FACS) and novel microfluidics, with the ultimate goal of producing sequencing libraries from single-amplified genomes (SAGs), with increased ability to target rare or novel microbes relative to "blind" sorting techniques.

(378e) High-throughput, Comprehensive Analysis of Eukaryotic Translation Initiation Motifs by FACS-Seq

Noderer, W., Stanford University
Wang, C., Stanford University
Originally presented on: 11/5/2013 16:27:00 - 16:45:00

The probability that a scanning ribosome will stop at a start codon and initiate translation is a function of the sequence surrounding the AUG. Unlike bacterial translation initiation, which uses well defined Watson-Crick base paring between the ribosome and the ribosome binding site, eukaryotic translation initiation is not well understood. Marilyn Kozak's consensus sequence provides helpful rules of thumb but is not quantitative. We have revisited this problem using a high-throughput, quantitative approach named FACS-seq which couples fluorescence activated cell sorting (FACS) with next-generation sequencing. Our mutational library consisted of random permutations of the 6 bases preceding and 2 bases following a start codon driving the expression of a fluorescence reporter. In agreement to Kozak's results, we find the -3 position to be the largest determinant of translation initiation. Contrary to Kozak's results, we find the +4 to be of little importance and instead find the two bases neighboring the -3 position to play a major role. A translation initiation prediction model incorporating the FACS-seq data was found to be in agreement with ribosomal footprinting data and was able to predict translation of upstream open reading frames. In principle, the FACS-seq method described here could be applied to any mutational library that can be paired with a fluorescence reporter.

(378f) Engineering Orthogonal Protein-Nucleic Acid Pairs for Cellular Circuits

Mansell, T. J., Johns Hopkins University
Lynch, S. A., University of Colorado at Boulder
Boyle, N. R., University of Colorado Boulder
Gill, R. T., University of Colorado Boulder
Originally presented on: 11/5/2013 16:45:00 - 17:03:00

Cellular circuits in synthetic biology often depend on inducible regulators like the *lac* and *tet* repressors. However, increased circuit complexity is currently limited by the availability of orthogonal regulators. We present a method for generation of directed libraries of regulators for orthogonal binding of DNA and RNA sequences in *E. coli*. Our unique selection-counterselection strategy allows for selection of repressors that do bind a target (i.e., novel) sequence but do not bind a pre-determined sequence of interest (e.g., wild type). We leverage the technologies of multiplex oligo synthesis and next-generation high-throughput sequencing to create libraries and glean sequence-activity relationship data from selections to iterate evolution in a rational, directed way.

(378g) Designing Synthetic Genes for Recombinant Expression in Synthetic Biotechnology

Chung, B., Bioprocessing Technology Institute (A*STAR)
Lee, D. Y., National University of Singapore
Originally presented on: 11/5/2013 17:03:00 - 17:21:00

Abstract

System-wide data, such as genomics, transcriptomics, proteomics and metabolomics, generated by high-throughput experimental techniques have fueled the rapid advancement in systems biology research over the past decade. This wealth of information generated in systems biology has equipped us with the knowledge to challenge the new frontier of synthetic biology. Hence, in this study, we demonstrate a methodology for employing omics data to computationally design synthetic genes with enhanced recombinant expression capabilities. In the biotechnology industry, recombinant DNA technology is commonly used to achieve high productivity of certain desired biochemical in host organisms such as *Escherichia coli*, *Saccharomyces cerevisiae*, *Pichia pastoris* and Chinese hamster ovary (CHO) cells. Accordingly, high heterologous expression levels of the recombinant genes in these host organisms are required. Nonetheless, these foreign genes are typically poorly expressed because they have not been evolved for efficient expression in the selected host organism. Consequently, molecular biology techniques have to be employed to either genetically engineer the host organism or modify the recombinant genes for enhanced heterologous protein expression. However, since the former approach is more challenging to implement and may adversely affect the host cells' physiology, the latter approach of redesigning the recombinant genes to generate synthetic genes is usually adopted. In this aspect, the issue of codon usage bias has been well-known to be a key factor affecting recombinant protein expressivity, more specifically at the step of translating mRNA to peptide chain (Gustafsson et al., 2004). Several codon optimization software tools, such as Gene Designer (Villalobos et al., 2006), JCat (Grote et al., 2005) and OPTIMIZER (Fuglsang, 2003), have been developed to generate synthetic genes with improved codon usage bias. Most of these earlier algorithms optimize the genetic sequence based on individual codon usage (ICU). Yet, the pairing of codons, sometimes referred to as codon context (CC), is also known to significantly affect mRNA translation efficiency (Hatfield and Roth, 2007; Coleman et al., 2008). Thus, it is relevant to incorporate both ICU and CC as design parameters under a multi-objective optimization framework to examine the relevant importance of these two parameters. Towards this end, we have developed a computational tool based on the Non-dominated Sorting Genetic Algorithm II (NSGA-II) to optimize the synthetic gene design with respect to both ICU and CC (Chung and Lee, 2012). This tool has been applied to optimize recombinant genes for heterologous expression in *P. pastoris* (Ahn et al., 2013) and CHO cells (Chung et al., 2013). The *in vivo*expression of the optimized synthetic genes in these two expression hosts has consistently revealed that CC-optimized genes were capable of higher expression levels than the ICU-optimized ones. This suggests that CC may be a more important design criterion than the widely considered ICU. Henceforth, the incorporation of the CC parameter into the codon optimization algorithm is expected to improve the efficacy of existing tools. In addition, the developed multi-objective codon optimization framework based on NSGA-II will be a good implementation strategy for future applications since it can be easily expanded to include more design parameters.

Acknowledgements

This work was supported by the National University of Singapore, Biomedical Research Council of A*STAR (Agency for Science, Technology and Research), Singapore, and a grant from the Next-Generation BioGreen 21 Program (SSAC, No. PJ009520), Rural Development Administration, Republic of Korea.

References

Ahn, J., Jang, M., Chung, B.K., Lee, D.Y., Lee, H., Jung, J.K., Choi, E.S., (2013) Comparison of relative importance of codon usage and codon-pair context as gene design parameters for improving protein expression in Pichia pastoris. *Submitted.*

Chung, B.K., Lee, D.Y., (2012) Computational codon optimization of synthetic gene for protein expression. *BMC Syst Biol*6, 134.

Chung, B.K., Yusufi, F.N.K., Mariati, Yang, Y., Lee, D.Y., (2013) Enhanced expression of codon optimized interferon gamma in CHO cells. *Submitted.*

Coleman, J.R., Papamichail, D., Skiena, S., Futcher, B., Wimmer, E., Mueller, S., (2008) Virus attenuation by genome-scale changes in codon pair bias. *Science*320, 1784-1787.

Fuglsang, A., (2003) Codon optimizer: a freeware tool for codon optimization. *Protein Expr Purif*31, 247-249.

Grote, A., Hiller, K., Scheer, M., Munch, R., Nortemann, B., Hempel, D.C., Jahn, D., (2005) JCat: a novel tool to adapt codon usage of a target gene to its potential expression host. *Nucleic Acids Res*33, W526-531.

Gustafsson, C., Govindarajan, S., Minshull, J., (2004) Codon bias and heterologous protein expression. *Trends Biotechnol*22, 346-353.

Hatfield, G.W., Roth, D.A., (2007) Optimizing scaleup yield for protein production: Computationally Optimized DNA Assembly (CODA) and Translation Engineering. *Biotechnol Annu Rev*13, 27-42.

Villalobos, A., Ness, J.E., Gustafsson, C., Minshull, J., Govindarajan, S., (2006) Gene Designer: a synthetic biology tool for constructing artificial DNA segments. *BMC Bioinformatics* 7, 285.

(378h) An Integrated Pipeline for Organism Engineering

Canton, B.,

Originally presented on: 11/5/2013 17:21:00 - 17:39:00

Ginkgo BioWorks is working to build a reliable and cost effective pipeline for engineering bacteria to produce fuels and high value chemicals. I will argue that there is no "silver bullet" technology to address this goal; instead, an integrated, balanced mix of technologies is the most plausible approach to making the engineering of biology more standard, more predictable, and hence more scalable. Our pipeline consists of tightly-integrated workflows for designing, building, and testing engineered organisms that each rely on a combination of software, hardware, and wetware. I will demonstrate how detailed data recording and analysis can enable the improvement and scaling of a novel DNA assembly scheme. Next, I will describe a high-throughput metabolomics workflow that captures a rich picture of the metabolic state of engineered cells. This metabolomics workflow depends on the latest hardware advances in mass spectrometry and custom software to process and analyze the large resultant datasets. Finally, I will show how we are deploying natural biological parts to control the behavior of engineered organisms in new ways.

(447a) Low Abundance of Respiratory Nitrate Reductase Is Essential for Escherichia Coli in Resistance to Aminoglycoside and Cephalosporin

Wang, S., Xiamen University
Ma, Y., Sun Yat-sen University
Guo, C., Sun Yat-sen University
Li, H.,
Peng, X.,
Originally presented on: 11/6/2013 8:30:00 - 8:48:00

The increasing incidence of antibiotic-resistant bacteria has posed a serious threat to human health and aquaculture today and thus has become a scientific issue worldwide. A line of evidence has indicated that four mechanisms contribute to the bacterial resistance, including modification or hydrolysis of enzymes, modification of drug targets, activation of efflux pump systems, and reduce of outer membrane permeability. Despite these progresses over past sixty years, our understanding of the resistant mechanisms is still incomplete, which should be one of the major causes resulting in antibiotic-resistant bacteria out of control. Thus, further investigation of bacterial resistant strategies to antibiotics may be necessary for combating antibiotic-resistant bacteria.

The recently developed technologies are expected to facilitate the study of bacterial antibiotic-resistant mechanisms. For example, protein-protein interactions, being important for the majority of biological functions including antibiotic resistance, may be potentiated using highly throughput screening methods of 2-D native PAGE (N-PAGE) and 2-D blue native PAGE (BN-PAGE) based proteomics. The two methods have been proven especially useful for investigation of native protein complexes in the context of proteomics. Comparatively, N-PAGE based on proteomics may offer more information on labile protein complexes containing metabolic enzymes than BN-PAGE based proteomics. Using the two approaches, researchers have investigated membrane and outer membrane protein complexes in *Escherichia coli* and identified a series of homogeneous and heterogeneous complexes. The identified membrane-related protein complexes may contain cytoplasmic and periplasmic proteins since approximately 30 - 40% *E. coli* proteins may function in the membrane of the cell envelope, in which there are soluble proteins tethered to the inner and outer membranes through hydrophobic patches, lipid moieties, or charge interactions or in membrane protein complexes. On the other hand, recent reports have indicated that metabolic enzymes are related to antibiotic resistance, and the regulation of metabolism elevates susceptibility of bacterial persisters to antibiotics. We have also shown that Na(+) NQR complex is essential for *Vibrio alginolyticus* in resistance to balofloxacin using N-PAGE based proteomics coupled with functional validation. We hypothesized that the comparative N-PAGE approach might provide metabolic-related protein complexes that contributed to antibiotic resistance for further functional investigation since many of metabolic enzymes are detected in membrane protein fraction. Here, N-PAGE based proteomics is used to identify a metabolic protein complex in membrane fraction of *Escherichia coli* K12 BW25113 and then the role of the complex in bacterial antibiotic resistance is further investigated.

Separation of membrane proteins and detection of Nar complex in *E. coli* K12 BW25113 membrane protein fraction using N-PAGE based proteomics. N-PAGE was used to separate the membrane proteins of *E. coli* K12 BW25113. Approximately a hundred of clear protein spots were observed on each of the 2-DE gels stained with cCBB R-250, most of which were located in the range of 25-120 kDa with minimal streaks. The streaks could be resulted from the diffusion of the same proteins during the first dimensional native-PAGE as reported previously. These well visualized protein spots were excised from 2-DE gels and analyzed by MALDI/TOF-TOF analysis. A hundred of proteins were successfully identified. They represented ninety-four unique proteins since five proteins had more than one spot. We interestingly found that two components, NarG and NarH, of Nar complex were lined, which did not show in our previous N-PAGE to detection of *E. coli* DH5α membrane protein fraction. Nar complex includes four polypeptides NarG, NarH, NarJ and NarI of molecular weight 138.7, 57.7, 26.5 and 25.5 kDa, respectively. NarJ and NarI did not appear in the gel probably due to the low molecular weight. NarG and NarH were further validated in 2-DE Western blotting. The complex is capable of reducing nitrate using normal physiological substrates and is clearly the major respiratory Nar in *E. coli* since it accounts for 98% of the Nar activity when fully induced. The result provides a chance to investigate a role of the complex contributed into bacterial energy metabolism in bacterial antibiotic resistance. In addition, sixteen out of the 94 proteins were validated using Western blotting. We identified the subunits A and B of the BAM complex in two distinct protein complexes. The BAM complex consists of one outer membrane protein, BamA (formerly YaeT), and four lipoproteins, BamB, BamC, BamD and BamE (formerly YfgL, NlpB, YfiO and SmpA, respectively). The other three were absence, which might be related to protein abundance, molecular weight and binding force. Further pathway analysis was carried out and obtained 90 annotation out of the 119 proteins identified. Among the 90 annotated proteins, 76 were enriched in 16 pathways. They are mainly involved in carbon metabolism, nitrogen metabolism, oxidative phosphorylation. Furthermore, we investigated the relationships between protein complexes containing altered abundance of protein(s) and antibiotic resistance.

Validation of Nar complex in *E. coli* K12 BW25113 membrane proteins. Then, the Nar complex was validated using two assays. First, far-Western blotting was used to investigate the interaction of NarH with NarG. Our results indicated that the bait protein was NarH when NarG was used as the prey protein. Then, the complex was further validated by N-PAGE analysis of *narH* and *narG* deletion mutants. This was based on the hypothesis that when one protein within a protein complex is absent, the others may disappear from N-PAGE gel. Our results showed that loss of *narG* caused disappearance of NarH, while deletion of *narH* resulted in loss of NarG in the same line, meaning that NarG-NarH complex existed in the line. These results indicate that Nar complex detected by the N-PAGE based proteomics is reliable, and thus may be used for investigation of antibiotic resistance.

Characterization of Nar complex in five antibiotic-resistant *E. coli* strains. In order to investigate whether abundance of Nar complex components was altered to resist antibiotics, the N-PAGE based proteomics approach was used to detect NarG and NarH levels in SM-R, GEN-R, CAZ-R, TET-R and NA-R. Significant changes were detected, showing that NarG and NarH disappeared in the SM-R, CAZ-R, TET-R and decreased in GEN-R and NA-R. These results not only indicate that Nar complex may play an important role in antibiotic resistance, but also show that the role is related the classes of antibiotics.

Functional characterization of *narG* and *narH* deletion mutants in response to five antibiotics. We next performed functional characterization of the Nar complex proteins using *narG* or *narH* deletion mutants. The performance was carried out using two antimicrobial susceptibility assays: MIC assay and survival capability assay. Deletion of *narG* or *narH* resulted in distinct elevation of MICs in medium with SM or GEN, and two-fold decrease of MICs in medium with TET, and no significant changes in medium with CAZ or NA. The survival capability assay was also carried out to further investigate functional characterization of the mutants due the assay is more sensitive than MIC assay. These bacteria including *ΔnarI* and *ΔnarJ* were cultured in tubes with series of two-fold dilution of SM, GEN, CAZ, TET and NA concentrations from 0.625 to 5 μg/mL, 0.1953-1.1719 μg/mL, 0.0781-0.3125 μg/mL, 0.78125-3.25 μg/mL, 1.25-7.5 μg/mL, respectively. The resulting cultures were assayed by measurement of optical density at 600 nm and the survival capability was determined by division of OD values of the cultures with an antibiotic over those without the drug. The survival rates of the mutants cultured in medium with SM, GEN or CAZ were significantly increased ($P<0.01$) in a dose-dependent manner. However, lower survival capabilities were detected in the mutants cultured in medium with TET and NA. Generally, the results obtained from survival capability assay were in consistence with those obtained from MIC assay. In summary, the results obtained from functional characterization of *ΔnarG* and *ΔnarH* were consistent with the changes of NarG and NarH at the protein abundance which was detected by 2-DE analysis of these antibiotic-resistant bacterial strains.

In summary, N-PAGE based proteomics approach was used to investigate *E. coli* K12 BW25113 membrane protein complexes and resulted in the detection of NarG and NarH, two main components of Nar complex. The decreased abundance of the two proteins was characterized in SM-R, GEN-R, CAZ-R, TET-R and NA-R. Functional tests using *narG*- and *narH*-deleted mutants demonstrated distinctly elevated resistance to aminoglycosides and CAZ. It is documented that low abundance of Nar complex is required for *E. coli* resistance to aminoglycosides and CAZ, which is the first report on that *E. coli* Nar metabolism alters bacterial resistance to antibiotics. This finding can serve to propose novel strategies for treating infections caused by aminoglycoside- and CAZ-resistant *E. coli* through activation of Nar activity.

ACKNOWLODGEMENTS

This work was sponsored by grants from NSFC projects (30972279, 40976080), Guangdong Provincial Science and technology projects (2012A031100004). Doctoral Fund of Ministry of Education of China (100171110029)

(447b) Cell Type-Specific Proteomic Profiling in Caenorhabditis Elegans

Yuet, K. P., MIT
Doma, M. K.,
Ngo, J. T.,
Sternberg, P. W., California Institute of Technology
Tirrell, D. A., California Institute of Technology
Originally presented on: 11/6/2013 8:48:00 - 9:06:00

In a complex eukaryote like *C. elegans*, cell and tissue heterogeneity restricts the usefulness of large-scale, mass spectrometry-based analysis of an organism's proteome. As worm cell lines are not available and enriching for specific cells or tissues is challenging, researchers cannot systematically identify low abundance proteins expressed in specific cells or tissues from whole-worm lysates. To isolate protein from specific cells, we have engineered *C. elegans* phenylalanyl-tRNA synthetases capable of appending azide- or alkynyl-bearing analogs of phenylalanine to $tRNA^{Phe}$. As these analogs are not substrates for any of the wild-type aminoacyl-tRNA synthetases in *C. elegans*, we achieved cell type selectivity by spatially restricting the expression of a mutant synthetase using cell type-specific promoters. Because proteins from those cells that express the mutant synthetase -- and only those cells -- contain the analogs, they can be enriched by treatment with alkyne- or azide-functionalized biotin reagents and subsequent affinity chromatography for identification by LC/MS-MS.

(447c) Phospho-Proteomics Reveals Oncogenic Phospho-Tyrosine Signaling Networks in Cancers Lacking Mutated Or Amplified Tyrosine Kinases

Graham, N. A., California Institute of Technology
Drake, J. M., University of California, Los Angeles
Tahmasian, M., University of California, Los Angeles
Pienta, K. J., University of Michigan, Ann Arbor
Witte, O. N., University of California, Los Angeles
Graeber, T. G., University of California, Los Angeles
Originally presented on: 11/6/2013 9:06:00 - 9:24:00

Introduction: Many human tumors exhibit aberrant activation of tyrosine kinase signaling due to mutation or DNA amplification of tyrosine kinases (eg, EGFR). In contrast, prostate cancer and some subtypes of glioblastoma (GBM) lack mutated or amplified tyrosine kinases. To investigate the regulation of phospho-tyrosine signaling in the absence of tyrosine kinase mutations or amplifications, we applied quantitative, label-free mass spectrometry to a GBM cell line, a mouse model of prostate cancer and human metastatic castration resistant prostate cancer (CRPC) biopsies.

Materials and Methods: Human GBM cells (U87) were treated with glucose starvation, epidermal growth factor (EGF), hydrogen peroxide (H_2O_2) or the phosphatase inhibitor vanadate. Mouse models of prostate cancer were generated by expressing combinations of non-tyrosine kinase oncogenes commonly found in prostate cancer (constitutively active AKT, androgen receptor (AR), ERG and K-Ras G12V) in the prostate *in vivo* regeneration model system (Lawson et al, 2009). Human treatment naÃ¯ve and metastatic castration resistant prostate cancer (CRPC) samples were obtained from the UCLA Translational Pathology Core Laboratory and the University of Michigan Rapid Autopsy Program. Quantitative, label-free mass spectrometry was performed on peptides immunoprecipitated using a pan-specific anti-phospho-tyrosine antibody (Drake et al, 2012; Graham et al, 2012).

Results and Discussion: U87 cells, which lack mutated or amplified tyrosine kinases, were treated with glucose starvation, EGF, H_2O_2 or vanadate. Each stimulus induced a unique signature of phospho-tyrosine signaling (Fig. 1a, n = 46 peptides). Unexpectedly, glucose starvation induced phosphorylation of proteins associated with focal adhesions, which suggested additional experiments that revealed a positive feedback loop between tyrosine kinase signaling and NADPH oxidase activity within focal adhesions (Graham et al, 2012). Next, analysis of mouse models of prostate cancer expressing non-tyrosine kinase oncogenes (eg, activated Akt and overexpressed ERG) (Drake et al, 2012) revealed oncogene-specific signatures of phospho-tyrosine signaling (Fig. 1b, n = 147 peptides), including an enrichment of EGFR substrates in Akt-ERG tumors. Finally, analysis of human prostate cancer patient samples demonstrated differential activation of tyrosine kinase signaling in treatment naÃ¯ve and metastatic CRPC tissues (Fig. 1c, n = 126 peptides), including the presence of Src kinase substrates in metastatic tumors.

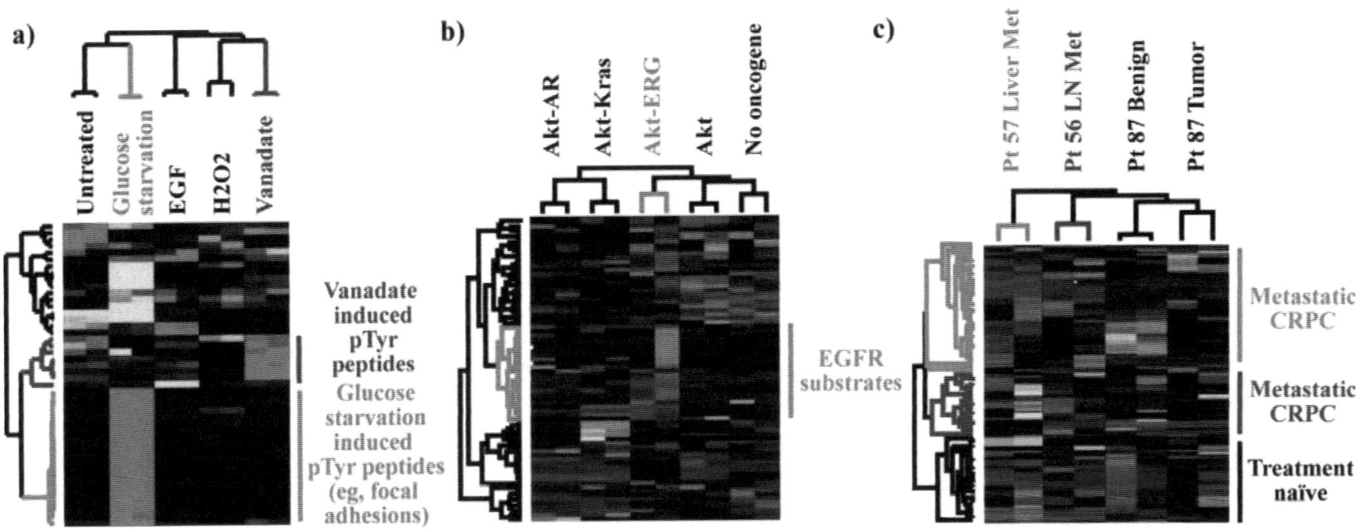

Figure 1. Hierarchical clustering of phospho-tyrosine peptides in a) U87 cells, b) mouse models of prostate cancer c) human CRPC patient (Pt) samples (LN, lymph node; Met, metastasis). Rows represent phospho-tyrosine peptides and columns represent samples run in duplicate. Red and green denote high and low phosphorylation, respectively.

Conclusions: We applied quantitative, label-free mass spectrometry to investigate regulation of phospho-tyrosine signaling in systems without mutated or amplified tyrosine kinases. This analysis revealed that a) metabolic perturbation (ie, glucose starvation) induces a positive feedback loop leading to focal adhesion-related tyrosine kinase signaling in U87 cells, b) the non-tyrosine kinase oncogenes Akt and ERG activate EGFR signaling in a mouse model of prostate cancer, and c) metastatic human CRPC biopsies exhibit elevated Src signaling compared to treatment naÃ¯ve biopsies. Together, these data demonstrate that cell lines and tumors lacking tyrosine kinase mutations or DNA amplifications can exhibit elevated tyrosine kinase signaling, suggesting that tyrosine kinase inhibitors may be effective therapeutics even in these cancers.

Acknowledgements: N.A.G. is supported by the UCLA Scholars in Oncologic Molecular Imaging (SOMI) program, NIH grant R25T CA098010. J.M.D. is supported by the Department of Defense Prostate Cancer Research Program W81XWH-11-1-0504.

References: Drake, J.M., *PNAS* (2012), 109(5), 1643-1648; Graham, N.A., et al, *Mol Syst Biol* (2012), 8:589; Lawson, D.A., et al, *PNAS* (2009).

(447d) Ovarian Cancer Metabolomic Profiles Differ in Ovarian Cancer Initiating Cells

Vermeersch, K. A., Georgia Institute of Technology
Styczynski, M. P., Georgia Institute of Technology
Originally presented on: 11/6/2013 9:24:00 - 9:42:00

A key emerging area in cancer research is the study of stem-like "cancer initiating cells" (CICs) in tumor populations. CICs are a major concern for effective cancer treatment, as they can self-renew, can differentiate into cancer cells, and exhibit chemoresistance; they are thus suspected as a primary cause of cancer recurrence. Very little is known about the molecular workings of these cells, and we must start to understand them in order to effectively treat CICs. Metabolism is widely known to be dysfunctional in cancer, most famously via the Warburg effect, but there is limited knowledge about CIC metabolism and about cancer metabolism as a whole. Metabolomics is the study of cellular metabolism as a whole by attempting to measure the levels of as many metabolites as possible. Metabolite levels integrate the effects of gene regulation, environmental perturbations, and numerous other levels of regulation.
By profiling the metabolism over time, we can obtain a systems-level understanding of the metabolic dynamics of CICs, which could profoundly affect our understanding, or even treatment, of cancer and CICs.

Here we present the results of an *in vitro* glucose deprivation experiment to model *in vivo* tumor conditions in ovarian cancer cells (OCCs) and ovarian CICs (OCICs). Extracellular and intracellular metabolite samples were taken over a period of two days and were analyzed using two dimensional gas chromatography – mass spectrometry (GCxGC-MS). To our knowledge, this is the first metabolic profiling of OCIC metabolism. Distinct metabolic profiles were found for the two cell types and between the control and glucose-deprived cells. Interestingly, independent of the differences imposed on the cells by glucose deprivation, there is a characteristic set of intracellular metabolites that only appear in OCCs, but not OCICs. These metabolites are spread across multiple pathways, suggesting system-wide metabolic alteration of OCICs from OCCs and potentially serving as indicators of CIC transition or even as therapeutic targets.

(447e) Integrated Energetics and Flux Analysis Reveals Differential Metabolic Reprogramming in Cancer Cells

Achreja, A., Indian Institute of Technology Madras
Yang, L.,
Nagrath, D., Rice University
Originally presented on: 11/6/2013 9:42:00 - 10:00:00

Cancer cell metabolism was investigated using systems biology tools such as metabolic flux analysis (MFA) to reveal intracellular cancer cell metabolic flux distribution. To obtain a complete role of nutrient-induced deprivation in metabolic reprogramming of cancer cells, we need to use metabolic flux analysis to determine unknown intracellular fluxes. Here, we have developed an integrated energy balance and flux balance analysis framework for analyzing cancer cell metabolism. Our framework allows estimation of intracellular fluxes and reveals specific adaptations and dysregulation of metabolic pathways in ovarian cancer cells as a result of the nutrient deprivation conditions. Our findings indicate that NAD+/NADH ratio regulates glutamine's direct effect on lactate production irrespective of the bioenergetic substrate requirement for different kinds of cancer cells. Our findings, using quantification of metabolites pertaining to central carbon and nitrogen metabolism provided clues as to which pathways relevant to glycolysis, the pentose phosphate pathway, the TCA, urea cycles and lipid cycles, amino acid and nucleotide metabolisms may be altered by nutrients in cancer cells. This resulted in identification of dominant factors which control cancer metabolism. We believe that development of our oncosecretomics based metabolic strategy provides an unprecedented opportunity for the development of nutritional supplementation that can deregulate the adaptations and alterations induced by various nutrients in cancer metabolism.

(447f) Temporal Metabolic Profiling of Plasma in Response to Endotoxemia in Humans

Kamisoglu, K., Rutgers, The State University of New Jersey
Sleight, K., Rutgers, The State University of New Jersey
Calvano, S. E., Robert Wood Johnson Medical School
Corbett, S. A., Robert Wood Johnson Medical School
Androulakis, I. P., Rutgers University
Originally presented on: 11/6/2013 10:00:00 - 10:18:00

Elective administration of bacterial endotoxin (LPS) to healthy human subjects has been used as a reproducible experimental procedure providing mechanistic insights into how cells, tissues and organs respond to systemic inflammation. Low doses of LPS transiently alter many physiologic and metabolic processes in a qualitatively similar manner to those observed after acute injury and systemic inflammation (1, 2); thus allow the analysis of body's responses to infectious stress at many physiological levels for building better understanding of the relevant pathology. Analysis of the metabolic response, in this regard, is of special interest since metabolic composition of a tissue is uniquely altered in response to stimuli due to collective effects of the regulations at transcriptional and translational levels. Therefore, concentrations of metabolites of a sample at a given time, i.e. the ?metabolome? (3), can be thought of as the metabolic fingerprint representative of the state of the body at that time and give information about the dominant regulatory mechanisms. Although the changes in the major metabolites, such as lipids, amino acids and glucose, has been previously documented for human endotoxemia (4); current study constitutes the first attempt of a complete metabonomic analysis that describes the coherent temporal patterns of the metabolic landscape in plasma following exposure to LPS.

Metabonomic analysis is comprised of global biochemical profiles determined in human plasma samples from healthy subjects. Nineteen subjects were administered either placebo (saline, n=4) or LPS (n=15) by a bolus injection and blood samples were collected at 5 time points within a 24 hour post-treatment period. Global biochemical profiles obtained by GC/MS and LC/MS/MS platforms represented information on 366 biochemicals including amino acids, short peptides, carbohydrates, lipids, nucleotides, cofactors and vitamins, xenobiotics and intermediate products of major energy production pathways. Within this dataset, we first identified those metabolites which have differential temporal profiles between control and LPS groups. The most significant difference between the two groups was at 6hr time point, which was also an inflection point separating development and recovery phases of the LPS induced metabolic changes. Then, through consensus clustering (5), we identified subsets of the metabolites with common coherent profiles. This analysis yielded two clusters with opposing directionality as shown in Figure 1. The first cluster (16 metabolites) was up-regulated within the first 6hr and down-regulated by the 24th hr and was mostly composed of metabolites from pathways related to lipid metabolism. The second cluster (21 metabolites), in contrast, was down-regulated within the first 6hr post-LPS, and then up-regulated by the 24th hr. Strikingly

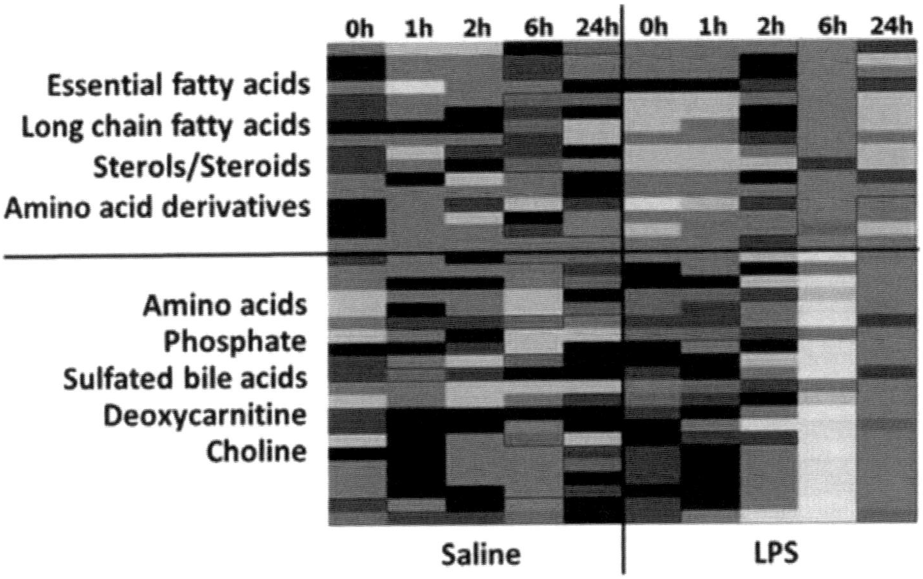

Figure 1: Differential patterns of metabolic response to LPS. Two clusters of plasma metabolites reflect two distinct patterns with opposing temporal directionality.

Preferential early clearance of amino acids can be attributed to hepatic uptake for the synthesis of acute phase proteins, while build-up of fatty acids in the plasma can be due to extensive lipolysis as a result of cytokine-mediated changes in lipid metabolism. Also, early build-up of some precursor molecules in sterol/steroid related pathways may indicate the propensity for hormone biosynthesis to suppress inflammation and promote recovery. These results highlight the changes in lipid and amino acid metabolism in response to infectious stress while also reflecting the most-closely associated plasma metabolites pointing out to co-regulatory schemes governing that response.

References

1. Lowry SF. Human endotoxemia: a model for mechanistic insight and therapeutic targeting. Shock. 2005;24 Suppl 1:94-100.

2. Calvano SE, Coyle SM. Experimental human endotoxemia: a model of the systemic inflammatory response syndrome? Surg Infect (Larchmt). 2012;13(5):293-9. PMCID: 3503465.

3. Nicholson JK, Lindon JC. Systems biology: Metabonomics. Nature. 2008;455(7216):1054-6.

4. Fong YM, Marano MA, Moldawer LL, Wei H, Calvano SE, Kenney JS, et al. The acute splanchnic and peripheral tissue metabolic response to endotoxin in humans. J Clin Invest. 1990;85(6):1896-904.

5. Nguyen TT, Nowakowski RS, Androulakis IP. Unsupervised selection of highly coexpressed and noncoexpressed genes using a consensus clustering approach. OMICS A Journal of Integrative Biology. 2009;13(3):219-37.

(447g) Metabolic Regulation for Double Increasing Balofloxacin Concentrations in Vibrio Alginolyticus

Peng, X.,
Yang, M., Tibtet vocational technical college
Zhao, X.,
Wang, S.,
Tian, Y.,
Guo, C.,
Liu, X.,
Li, H.,

Originally presented on: 11/6/2013 10:18:00 - 10:36:00

Metabolome refers to the complete set of small-molecule metabolites to be found within a biological sample. Metabolomics approach provides the global qualitative and quantitative analysis of the small-molecule metabolites and shows great promise as a mean to identify biomarkers of drug efficacy and understand metabolic mechanisms. In the present study, metabolomics carrying cellular metabolites of *Vibrio alginolyticus* has been developed to analyze dynamic changes in metabolites induced upon stress with double increasing concentrations of balofloxacin (BLFX). The objective was twofold. First, we were interested in identifying metabolites that, being specifically regulated in response to BLFX exposure, might be involved in BLFX-stressed biomarkers. Second, we wanted to exploit the metabolomics data to identify, among the BLFX-stressed metabolites, compounds capable of being reasonable proofs describing the characteristic feature of *V. alginolyticus* in response to the double increasing concentrations of BLFX. We hypothesize that the achievement of the two objectives will tell us a new story on bacterial metabolic response to gradient increasing concentrations of antibiotic.

Effect of double increasing concentrations of BLFX on bacterial growth. *alginolyticus* cells were cultured in LB medium separately with 0, 1/64, 1/32, 1/16, 1/8, 1/4 and 1/2 MIC of BLFX. These cultures were harvested when the bacterial cell density in medium without the drug arrived at approximately 1.0 OD. The resulting OD was negatively related to the double increasing concentrations of BLFX. Compared with the BLFX-con0, bacterial survival rates were 98.5 ± 1.34, 97.2 ± 0.78, 95.0 ±0.39, 92.0±0.61, 87.8 ± 0.54, 83.4 ± 0.48 in BLFX-con64, BLFX-con32, BLFX-con16, BLFX-con8, BLFX-con4 and BLFX-2, respectively. Growth ability of the bacterium was a negative correlation to the double increasing concentrations of the drug in medium. Significant difference was found at $P < 0.01$ between them. These results indicate an association of the inhibitory role in bacterial growth with BLFX concentrations.

Characteristics of dynamically metabolic response to double increasing concentrations of BLFX. To investigate a differential metabolome of *V. alginolyticus* in response to BLFX, GC/MS based metabolomics approach was used to identify metabolic responses to the double increasing concentrations of BLFX. Quantitative data throughout the entire time course were obtained for 89 compounds from 128 peaks after deletion of nine peaks that were below the limit of quantitation or too unstable to yield reliable information. OPLS-DA approach distributed the seven sample groups into different levels with the double increasing concentrations of BLFX in 3-D scatter plot, in which BLFX-con0 and BLFX-con64 were located in the bottom face and in the interface between the bottom and lateral faces, respectively. The others were clearly separated from BLFX-con0 and the separation was trended with the double increasing BLFX concentrations in the lateral face. Dynamics of the 89 compounds was shown in a clustered heat-map format. The format showed that BLFX-con0, BLFX-con64 and BLFX-con32, BLFX-con16, BLFX-con8 and BLFX-con4, and single BLFX-con2 were clustered together. In the first two clusters, there was significant distance between BLFX-con0 and the other two, and between BLFX-con4 and the other two, indicating the quantitative/qualitative pattern of metabolic changes in response to the double increasing BLFX concentrations. These results show that the metabolic profiles have significantly adjusted as a result of the increased antibiotic stress.

Identification of dynamically differential metabolites in response to double increasing concentrations of BLFX. For determining the characteristic compounds in BLFX, the multi-class unpaired method of SAM was used to identify the metabolites whose concentration differed significantly among all samples collected across the double increasing concentrations of BLFX. Despite these groups being only double-fold elevation of BLFX concentrations, they could be differentiated from SAM analyses. SAM identified 47, 64, 67, 56, 59 and 78 peaks representing 35, 48, 47, 48, 46 and 53 metabolites in BLFX-con64, BLFX-con32, BLFX-con16, BLFX-con8, BLFX-con4 and BLFX-2, respectively. The 44 positively significant metabolites and 9 negatively significant metabolites from 78 peaks detected in BLFX-2, containing the all changes occurred in the other samples, in order of decreasing significance palmitelaidic acid, phosphoric acid, tetradecanoic acid, adenosine, morphinan, oleic acid as the first six metabolites. Interestingly, ratios of metabolite areas with BLFX doses were not exactly proportional in every concentration for majority of the metabolites with differential levels, although changing tendency of the differential metabolites was detected. These results indicate that BLFX-stressed *V. alginolyticus* showed a quantitative/qualitative change in the response.

Then, PCA was separately used to identify significant di erences in the data set. The first six metabolites from factor weight were similar in the seven comparisons in order of decreasing significance L-proline, butanedioic acid, silanamine, phosphoric acid, L-alanine, glycine.

Potential markers of interest were further extracted from S-plots constructed following analysis with OPLS-DA. Fourteen biomarkers were identified from the loading S-plots for metabolic pathway analysis since they showed statistically significant ($P < 0.05$) difference in abundance. Out of them, the first eight existed only in comparison cross the seven groups (BLFX-con all), in which negative regulation of phosphoric acid, palmitelaidic acid, glutamine, glycine, 5'-adenylic acid and 9H-purin-6-amine and positive regulation of butanedioic acid, L-proline, adenosine were characterized. VIP analysis confirmed the reliability of the first four biomarkers phosphoric acid, butanedioic acid, L-proline and palmitelaidic acid in the total comparison. The other four were validated in comparison between two groups. However, the regulation by each of these compounds did not run through whole responsive process to the double increasing BLFX concentrations. They showed an alternating renewal process of positive/negative regulation. For example, negative and positive regulations were detected in BLFX-con64 vs -con0 (-0.18), BLFX-con16 vs –con32 (-0.19) and BLFX-con32 vs –con64 (0.20), BLFX-con8 vs –con16 (0.53), BLFX-con4 vs –con8 (0.66), BLFX-con2 vs –con4 (0.86) for phosphoric acid, respectively. Generally, an alternating renewal positive/negative regulation was detected in first four comparisons, BLFX-con64 vs -con0, BLFX-con32 vs –con64, BLFX-con16 vs –con32 and BLFX-con8 vs –con16, and then changing tendency kept until the last comparison BLFX-con2 vs –con4 during the stress processes for each of the key metabolites. There were two change points, BLFX-con64 vs -con0, BLFX-con16 vs –con32, in the response process. These results obtained from S-plot were consistent with the finding obtained from the clustering. These results not only indicate the importance in the structure of advance by degrees for details, but also show the quantitative/qualitative pattern of metabolic changes in response to the double increasing concentrations of BLFX.

Identification of differential metabolic pathways in response to double increasing concentrations of BLFX. Furthermore, ingenuity network analysis was used to perform more detailed analysis of pathways and networks influenced by the double increasing concentrations of BLFX. Out of the significantly altered 77 metabolites, 66 were used for the analysis in deletion of 11 unclear pathway metabolites. To identify possible pathways that were affected by BLFX, metabolites contributing to the separation of control and groups with the double increasing BLFX concentrations were analyzed in structure of advance by degrees using MetPA. The analysis revealed metabolites, which were identified together were important for the bacterial response to the drug, were responsible for 27-39 pathways. When the impact-value threshold was set to 0.10, eight potential targets pathways were filtered out. They were inositol phosphate metabolism, glycine, serine and threonine metabolism, alanine, aspartate and glutamate metabolism, pantothenate and CoA biosynthesis, arginine and proline metabolism, glyoxylate and dicarboxylate metabolism, tryptophan metabolism, methane metabolism. However, the Raw p, the original p value calculated from the enrichment analysis, analysis showed an alternating renewal process of significant/insignificant changes in the double increasing concentrations of BLFX. For example, significant difference was detected in BLFX-con64 (0.001), BLFX-con32 (0.03) and BLFX-con8 (0.02), but not in BLFX-con16 (0.09), BLFX-con4 (0.32), BLFX-con2 (0.16) for glycine, serine and threonine metabolism (Table 3), suggesting an efficient strategy for the bacterium in response to BLFX through alternating renewal modification of pathways.

Identification of promoting constructive metabolism in response to double increasing BLFX concentrations. We further detected bacterial dry weight and protein content for understanding whether the bacterium response to BLFX mainly through promoting constructive metabolism or catabolism. First, bacterial dry weight was measured. An increasing tendency with the double increasing BLFX concentrations was detected with correlation coefficient ($P < 0.05$). Approximately 10 – 20 % dry weight was increased in medium with 1/8 – 1/2 MIC of BLFX. Then, significantly elevated total protein content was determined at $P < 0.01$ with a positive correlation between the double increasing BLFX concentrations and protein content. These results indicate that *V. alginolyticus* enhances protein constructive metabolism in response to BLFX.

In summary, our findings indicate that *V. alginolyticus* mounts an alternating renewal positive/negative regulation of metabolites and discontinuity modification of pathways for the increasing BLFX concentrations, forming characteristic changes of the quantitative/qualitative pattern. These findings have significant implication in controlling BLFX-resistant *V. alginolyticus* by adjusting metabolites and metabolic pathways.

This work was sponsored by grants from "973" project (2012CB114406) and NSFC projects (30972279, 40976080). Doctoral Fund of Ministry of Education of China (100171110029).

(447h) Metabolic Analysis of Saccharomyces Cerevisiae to Evaluate the Relationship Between Nitrogen Utilization Efficiency and Ethanol Tolerance During Alcoholic Fermentation

Lozada Contreras, M., University of California, Davis
Block, D. E., University of California Davis
Originally presented on: 11/6/2013 10:36:00 - 10:54:00

During alcoholic fermentation, the sugar utilization activity of yeast cells decreases in the presence of alcohol, followed eventually by a loss in cell viability. Understanding ethanol tolerance should allow its modulation in yeast strains with other desirable characteristics. Commercial yeast strains, given an identical initial nitrogen concentration and composition, will form different concentrations of biomass. Theoretically, strains that use nitrogen more efficiently to create more biomass will be able to tolerate higher levels of ethanol produced. The goal of this project is to understand the relationship between nutrient environment, nitrogen metabolism, and specific growth rate in *Saccharomyces cerevisiae* on a fundamental level, in order to develop a rational approach to strain modification for increasing ethanol tolerance. This will be achieved by determining the extracellular and intracellular concentrations of key metabolites in yeast collected during the exponential and stationary phases. Thirty-two commercial yeast strains were compiled with a wide range of growth and ethanol tolerance attributes. Nitrogen utilization efficiency and fermentation kinetic parameters were studied in each strain. Maximum optical densities for these yeast ranged from 2.68 to 6.17. A complete metabolic analysis will be performed to investigate the differences in carbon and nitrogen metabolism using supernatant and cell samples saved from fermentations using GC-MS.

(490a) Mathematical Modeling of Light-Mediated Cortisol Secretion and Evaluation of Downstream Effects On Cytokine Production

Mavroudis, P. D., Rutgers University
Corbett, S. A., Robert Wood Johnson Medical School
Calvano, S. E., UMDNJ-Robert Wood Johnson Medical School
Androulakis, I. P., Rutgers University
Originally presented on: 11/6/2013 12:30:00 - 12:48:00

Circadian rhythmicity in mammals is regulated by a light sensitive central clock in the suprachiasmatic nucleus (SCN) of the hypothalamus. Multisynaptic neural pathways from the SCN communicate photic information to other brain regions such as the paraventricular nucleus of hypothalamus with its ultimate release of corticotrophin releasing hormone (CRH) (3). In addition to the SCN and the circadian clocks residing in brain region (central clocks), most peripheral tissues express circadian rhythmicity driven in cellular level by autoregulatory feedback loops of clock genes and proteins (peripheral clocks). Cortisol is an important entrainer of peripheral clocks since its production and secretion is regulated by the hypothalamic-pituitary-adrenal (HPA) axis through the release of CRH from the hypothalamus and adrenocorticotrophin hormone (ACTH) from the anterior pituitary gland (11) and therefore is assumed to retain the photic information. Cortisol, apart from entraining peripheral oscillators, feeds back to hypothalamus and anterior pituitary inhibiting the secretion of CRH and ACTH forming in such a way a negative feedback that leads to cortisol oscillating behavior (9). It turns out that maintaining synchrony between the environmental light cycles, centrally driven hormones, and peripheral oscillators, is important for normal physiological function since its loss, demonstrated by a loss of phase coherence between physiological parameters (i.e. cortisol), is correlated with several severe pathologies (6, 8). Interestingly, there are several works showing a direct link between loss of light entrainment and deteriorating inflammatory responses both at the systemic level (i.e. fever, anorexia) as well as at molecular level (i.e. secretion of pro-inflammatory cytokines) (2).

In this work, we discuss the development of a mathematical model which aims to evaluate the dynamic relations between light entrainment of the HPA axis and its downstream effects on peripheral clocks. In particular, we aim to investigate the changes of cortisol circadian secretion as a result of altered photoperiods as well as different light stimuli throughout the day cycle, that ultimately lead to different circadian characteristics of inflammatory mediators (pro-inflammatory cytokine) secreted by peripheral cells.

To study the effects of photoperiod on cortisol secretion and ultimately on inflammatory markers, we propose a unified model that incorporates two compartments; a central oscillator simulating the cortisol circadian production from HPA axis, and a peripheral oscillator simulating clock gene regulation of pro- and anti-inflammatory cytokine circadian secretion. Centrally, oscillations are maintained by the negative feedback between cortisol and CRH/ACTH hormones (10), and peripherally by the negative feedback of PER and CRY clock proteins on the translocation of CLOCK/BMAL heterocomplex to the nucleus of peripheral immune cells (1). In accordance with previous mathematical models (4), we represent light entrainment of CRH compartment by including an additive term in CRH equation, that represents the neural mediated regulation of the paraventricular nucleus of the hypothalamus. Entrainment at the central level, results in a phase shift of cortisol secretion in order to be orchestrated with light cycles, that further lead to a phase change of peripheral clocks since cortisol entrains clock genes by binding to a glucocorticoid responsive element (GRE) at the promoter region of Per1 and Per2 genes (7, 12).

Model predictions demonstrate that altered light periods, throughout a 24hr light/dark cycle, induce altered phases both in the central compartment as well as in peripheral clocks. Ultimately, these changes lead to modified homeostatic secretion of pro-inflammatory cytokines thus impacting the immune response of the host. Upon different light schedules, the organism seems to be exposed to different levels of inflammatory mediators in accordance with what has been observed experimentally in rodent experiments (2).

Our model highlights the implications of linking light mediated cortisol secretion and peripheral circadian regulation of inflammatory cytokines. Our aim is to provide further understanding of the chronobiological effects of loss of environmental entrainment ultimately leading in many circumstances to deteriorating clinical outcomes (5).

References:

1. **Becker-Weimann S, Wolf J, Herzel H, and Kramer A**. Modeling feedback loops of the Mammalian circadian oscillator. *Biophys J* 87: 3023-3034, 2004.

2. **Bilbo SD, Drazen DL, Quan N, He L, and Nelson RJ**. Short day lengths attenuate the symptoms of infection in Siberian hamsters. *Proc Biol Sci* 269: 447-454, 2002.

3. **Buckley TM, and Schatzberg AF**. On the interactions of the hypothalamic-pituitary-adrenal (HPA) axis and sleep: normal HPA axis activity and circadian rhythm, exemplary sleep disorders. *J Clin Endocrinol Metab* 90: 3106-3114, 2005.

4. **Geier F, Becker-Weimann S, Kramer A, and Herzel H**. Entrainment in a model of the mammalian circadian oscillator. *J Biol Rhythms* 20: 83-93, 2005.

5. **Lowry SF**. The stressed host response to infection: the disruptive signals and rhythms of systemic inflammation. *Surg Clin North Am* 89: 311-326, vii, 2009.

6. **Mahoney MM**. Shift work, jet lag, and female reproduction. *Int J Endocrinol* 2010: 813764, 2010.

7. **Mavroudis PD, Scheff JD, Calvano SE, Lowry SF, and Androulakis IP**. Entrainment of peripheral clock genes by cortisol. *Physiol Genomics* 44: 607-621, 2012.

8. **Schernhammer ES, Laden F, Speizer FE, Willett WC, Hunter DJ, Kawachi I, and Colditz GA**. Rotating night shifts and risk of breast cancer in women participating in the nurses' health study. *J Natl Cancer Inst* 93: 1563-1568, 2001.

9. **Spencer RL, Kalman BA, and Dhabhar FS**. Role of Endogenous Glucocorticoids in Immune System Function: Regulation and Counterregulation. In: *Comprehensive Physiology*John Wiley & Sons, Inc., 2001.

10. **Sriram K, Rodriguez-Fernandez M, and Doyle FJ, 3rd**. Modeling cortisol dynamics in the neuro-endocrine axis distinguishes normal, depression, and post-traumatic stress disorder (PTSD) in humans. *PLoS Comput Biol* 8: e1002379, 2012.

11. **Touitou Y, Haus E, Angeli A, Gatti G, and Masera R**. Chronobiology of the Hypothalamic-Pituitary-Adrenal and Renin-Angiotensin-Aldosterone Systems. In: *Biologic Rhythms in Clinical and Laboratory Medicine*Springer Berlin Heidelberg, 1992, p. 292-314.

12. **Yamamoto T, Nakahata Y, Tanaka M, Yoshida M, Soma H, Shinohara K, Yasuda A, Mamine T, and Takumi T**. Acute physical stress elevates mouse period1 mRNA expression in mouse peripheral tissues via a glucocorticoid-responsive element. *J Biol Chem* 280: 42036-42043, 2005.

(490b) The Src-Family Kinase Inhibitor PP2 Rescues Inducible Differentiation Events in Emergent Retinoic Acid-Resistant Myeloblastic Leukemia Cells

Jensen, H.,
Tasseff, R., Cornell University
Yen, A.,
Varner, J., Cornell University
Originally presented on: 11/6/2013 12:48:00 - 13:06:00

Retinoic acid is an embryonic morphogen and dietary factor that demonstrates chemotherapeutic efficacy in inducing maturation in leukemia cells. Using HL60 model human myeloid leukemia cells, where all-trans retinoic acid (RA) induces granulocytic differentiation, we developed two emergent RA-resistant HL60 cell lines which are characterized by loss of RA- inducible G1/G0 arrest, CD11b expression, inducible oxidative metabolism and p47phox expression. However, RA-treated RA- resistant HL60 continue to exhibit sustained MEK/ERK activation, and one of the two sequentially emergent resistant lines retains RA-inducible CD38 expression. Other signaling events that define the wild-type (WT) response are compromised, including c-Raf phosphorylation and increased expression of c-Cbl, Vav1, and the Src-family kinases (SFKs) Lyn and Fgr. As shown previously in WT HL60 cells, we found that the SFK inhibitor PP2 significantly increases G1/G0 cell cycle arrest, CD38 and CD11b expression, c-Raf phosphorylation and expression of the aforementioned regulators in RA-resistant HL60. The resistant cells were potentially incapable of developing inducible oxidative metabolism. These results motivate the concept that RA resistance can occur in steps, wherein growth arrest and other differentiation events may be recovered in both emergent lines. Investigating the mechanistic anomalies in resistant cell lines is of therapeutic significance and helps to mechanistically understand the response to retinoic acid's biological effects in WT HL60 cells.

(490c) Systemic Analysis of a Cellular Network Model of Liver Regeneration Reveals Potential Mechanisms of Regeneration Suppression

Cook, D., University of Delaware
Correnti, J., Thomas Jefferson University
Hoek, J., Thomas Jefferson University
Vadigepalli, R., Thomas Jefferson University
Ogunnaike, B. A., University of Delaware
Originally presented on: 11/6/2013 13:06:00 - 13:24:00

As the body's main detoxifying organ, the liver is routinely exposed to high levels of toxic compounds. Mature liver cells therefore retain the unique ability to enter the cell cycle and regenerate lost mass following even severe injury without impairing liver function. This tightly controlled response to an experimentally repeatable stress gives an ideal process to study the dynamic function of the organ. Up to a certain point, the liver's regulatory network is able to compensate for imbalances in strength and altered timing of factors; this is exemplified by animal-to-animal variability in gene and protein expression. However, after some threshold or in some imbalanced conditions, the network is no longer able to compensate for de-regulation and regeneration is impaired. We have developed a computational model of liver regeneration which allows for investigation into how the timing and balance of factors should respond in healthy animals and how they can be de-regulated to suppress regeneration.

A systems approach was taken to develop a computational model of cellular crosstalk during liver regeneration by integrating critical cellular interactions between non-parenchymal cells and hepatocytes identified from an extensive review of existing literature. Model parameters were fit using high-throughput protein measurements taken from wild-type mice over a time-course of the priming phase of regeneration following partial hepatecomy and from published data describing regeneration profiles. Local and global sensitivity analyses were performed to identify key model parameters affecting regeneration. Subsequent analysis of the model, focusing on those factors, revealed several principles governing the suppression of regeneration following partial hepatectomy. De-regulating pro-proliferative signals (by either reducing their relative strength or delaying their production) causes a delay in the initiation of regeneration, but ultimately no change in the total mass recovered. De-regulating the anti-proliferative signals (by either increasing their relative strength or increasing production rates) causes no changes to the initiation of regeneration, but leads ultimately to a decrease in total liver mass recovered. Inhibition of either pathway by itself is not enough to completely suppress regeneration; however, if both pathways are simultaneously de-regulated, regeneration can be completely suppressed. Additionally, the model implicates TGF-β produced by stellate cells along with decreased metabolic load as key factors causing the decreased regenerative response seen in one-lobe partial hepatectomies.

This work is a key piece of parallel efforts within our lab to identify cell-type specific, intracellular regulatory network models from high-throughput gene expression, miRNA expression, transcription factor binding, and protein levels. Future work to iteratively improve this model will include experimental investigation of model predictions and integration of intracellular transcriptional models into the framework of the cellular network model.

Research Support: R01 18873, T32 07463

(490d) Developing An Information Theoretic Framework for Model Selection in Systems Biology

DeVilbiss, F. T.,
Song, H. S., Purdue University
Ramkrishna, D., Purdue University
Originally presented on: 11/6/2013 13:24:00 - 13:42:00

In the process of model development, systems biologists often arrive at crossroads when it comes to selecting a specific modeling approach out of many possible ones to describe the functionality of a given network. A variety of system descriptions are available for such applications from low level resolutions to that of the organism. While there is no universal rule to determine the absolute utility of a model, certain metrics demonstrate promise in providing a coherent, rational, and objective basis for comparing model predictions. To address the problem of model selection, it is prudent to borrow arguments formulated from information theory. In this light, it becomes possible to distinguish models from one another in their ability to describe the regularity of a given system's output data. Capturing this regularity is in itself an act of compressing the data and therefore information theoretic methods are applicable. In this vein, one may apply the Minimum Description Length (MDL) principle to candidate biological models in order to compare how well these models describe regularity in data. Metrics derived upon MDL principles seek to identify a single model from a set of candidate models that best compresses features in data. This approach accomplishes this task through the consideration of both the model's likelihood of fit and inherent complexity. The model that best compresses data from a process can be said to be the most efficient and useful description of the system.

In this work, a handful of metrics developed upon these concepts, including Akaike Information Criteria (AIC), Bayesian Information Criterion (BIC), and Normalized Maximum Likelihood (NML) are considered for a group of dynamic metabolic models to determine which model best compresses metabolic data through maximizing model fit and minimizing model complexity. This analysis identifies a point of diminishing returns in which additional model complexity provides little gain in describing data accurately for both nested and non-nested metabolic models. Starting with a comparison of flux predictions made by constraint based, kinetic and cybernetic metabolic models, this work develops a framework that intends to be extended to other model comparison applications in systems biology.

(490e) Identification of Regulatory Structure and Kinetic Parameters of Biochemical Networks Via Mixed-Integer Dynamic Optimization

GuillÃ©n-GosÃ¡lbez, G., Universitat PolitÃ©cnica de Catalunya
Miro, A. Sr.,
Originally presented on: 11/6/2013 13:42:00 - 14:00:00

Abstract

Motivation:

The analysis of topological and kinetic information of biochemical networks from data obtained by high throughput techniques is a central topic in systems biology. Intensive researches have focused on estimating kinetic parameters from experimental observations in complete parameterized models. However, the problem of the simultaneous identification of the regulatory topology and model parameters from dynamic profiles has received much less attention to date.

Results:

We present a rigorous approach based on mathematical programming for the parameter estimation of models of biochemical networks that can predict the underlying regulatory structure as well. We formulate the task of identifying this structure along with the values of the kinetic parameters of the associated model as a mixed-integer dynamic optimization (MIDO) problem with two types of variables: (i) binary variables, used to model the existence of regulatory interactions and kinetic effects of metabolites in the network processes; and (ii) continuous variables, denoting metabolites concentrations and kinetic parameters values. The MIDO model seeks to optimize the Akaike criterion, which captures the trade-off between complexity (measured by the number of parameters) and accuracy (quantified by the least square error between experimental observations and *in silico* predictions) in an attempt to avoid overfitting to the extent possible. This MIDO is reformulated into a mixed-integer nonlinear program (MINLP) using orthogonal collocation on finite elements. The reformulated MINLP is solved via standard optimization tools in an iterative fashion in order to identify a set of plausible network topologies and associated kinetic parameters.

Conclusion:

The capabilities of our approach were tested in one benchmark problem. The example presented in this work, being simple, show that estimating parameters in dynamic kinetic models is far from being an easy task. Our algorithm was able to identify a set of plausible network topologies with their associated parameters that are consistent with the dynamic data available and that can be further refined using additional information and expert knowledge on the system.

(490f) Abstraction of Kinetic Models (AKM) Using Function and Parameter Co-Estimation Exploration

Hopkins, C., Tufts University
Hassoun, S., Tufts University
Originally presented on: 11/6/2013 14:00:00 - 14:18:00

Kinetic modeling attempts to describe the time-dependent behavior of every enzyme-catalyzed reaction in the network. Systems of ordinary differential equations express each reaction rate as a function of metabolite concentrations and rate constant parameters. The mechanistic knowledge of enzyme kinetics however is not always available. Furthermore, estimating a consistent set of rate parameters from time-series data requires a large experimental effort for even a moderately-sized network. This work develops approximate kinetic expressions for a module (*connected groups of reactions*) instead of developing kinetic expressions for *each*reaction within the module. The approximate kinetic expressions are thus derived only for the exchange metabolites, and not for the internal (non-exchange) metabolites or reactions. The two relevant questions are: (1) Does this approach yield a predictive module model? (2) Which metabolite concentrations (and measurements) can be eliminated when constructing such a model?

To explore the first question, we develop a systematic method for creating and evaluating module expressions, we refer to as ABK (Abstraction of Kinetic Models). We assume expressions will have the format of convenience kinetics [1], where each metabolite can influence another, by acting as either a reactant, product, activator, or inhibitor. Initially, we assume all metabolites are of interest and calculate kinetic expressions for each, tuning the expression parameters using a genetic algorithm. Next, we eliminate a metabolite from consideration, and recalculate all kinetic expressions, ranking resulting models based on smallest mean square error between measured and computed data points for exchange metabolites. Exhaustive elimination of all metabolite combinations is not computationally feasible. We thus use selective elimination, successively reducing the number of internal metabolites used in the expressions, creating new solutions.

The method was applied to two case studies. The first case study consists of reactions catalyzed by three glycolytic enzymes in *E. coli* (Glk, Pgi, and PfkA) for which detailed kinetic parameters have been determined. The second case study consists of reactions carrying mass flow and regulated through complex mechanistic interactions. The ABK algorithm generates a number of solutions, reported as a Pareto front, that trade prediction accuracy for complexity (number of metabolites used in the expressions). The benefits of reduced kinetic models are twofold: (1) reducing the number of measurements required for fitting the model, and (2) the resulting number of mathematical expressions is smaller compared to the number of equations needed to describe every reaction of the original network with equal detail. The presented method and the case studies show that expression simplification is possible while maintaining an overall predictive model. Our reduced models are suitable for building fast and predictive genome-scale kinetic models.

[1] W. Liebermeister and E. Klipp. (2006) Theor Biol Med Model 3: 41.

(490g) Reconstruction of Glycosylation Reaction Networks: Integration of Glycomics and Enzyme Data With Computer Models

Liu, G., State University of New York
Neelamegham, S., State University of New York
Originally presented on: 11/6/2013 14:18:00 - 14:36:00

Introduction: Glycosylation is an important post-translational modification that alters more than 50% of mammalian secreted and cell surface proteins [1]. Glycosylated proteins play structural and functional roles in diverse biological processes including inflammation, cancer and development. The recent development of advanced analytical tools, especially mass spectrometry (MS) methods, has heralded the emergence of Glycomics, a field where entire glycomes of cell systems are characterized. These data are starting to be stored in glycosylation specific databases like the Consortium for Functional Glycomics (CFG) Glycan Profiling database (http://www.functionalglycomics.org/glycomics/publicdata/glycoprofiling-new.jsp). It is currently challenging to quantitatively analyze MS data available in such databases in terms of identifying the underlying biosynthetic pathways that result in a particular MS profile. In this regard, since glycosyltransferase enzyme specificity and concentration data are not considered during conventional analysis, a direct linkage between enzyme activity changes to cell phenotype alterations is missing. To address this shortcoming, we developed a computational methodology to reconstruct species-specific and condition-specific cellular glycosylation pathways by integrating glycan structure information from MS data with enzyme-specific information.

Materials and Methods: An application package called GNAT (Glycosylation Network Analysis Toolbox) was developed in order to link MS data to biochemical data available in various databases and *in silico* glycosylation reaction network models.

Results and Discussion: Three tasks were undertaken. First, glycan information were gathered from MS experiment using GlycoWorkbench [2] to process raw data, annotate structures and calculate distributions. The enzyme-specific information were queried using enzyme databases, including BRENDA [3] and CAZY [4]. The collated data include information on glycan species, kinetic rate constants and enzyme substrate specificity. Second, a network construction algorithm was implemented to build a complete reaction network given the input glycan structures and species-specific enzyme data above. The inference algorithm developed is iterative, with each step using glycosyltransferase or glycosidase reaction mechanisms to either add or remove a monosaccharide residue from a given substrate. The synthesized network that results from the connection of all input MS glycan structures to pre-defined starting material is called the "master pathway". Third, subset network modeling, hierarchical clustering, principal component analysis and sensitivity analysis were applied to refine the reaction network. In this step, model predictions were compared with MS data to iteratively refine the model. The novel methods described above are implemented in our open-source MATLAB Toolbox, GNAT. The new features described above were tested using two case studies that relate to: i) N-Glycan processing in Chinese hamster ovary (CHO) cell and its mutants [5]; ii) O-glycosylation networks in human promyelocytic leukemia (HL60) cell [6]. These rigorous tests confirm that our approach can be used to construct computer models that quantitatively describe experimentally observed cell-type specific glycosylation reaction networks [7].

Conclusions: GNAT is enriched with new classes and functions to handle MS data, enzyme information, network inference, subset network generation and sensitivity analysis. By integrating glycomics and enzyme database information with modeling, the methods developed enable the synthesis of glycosylation network models that explain the MS data both qualitatively and quantitatively.

References:

1.Neelamegham, S., et al. Glycobiology, 2011. **21**(12): p. 1541-53.

2.Ceroni, A., et al. J Proteome Res, 2008. **7**(4): p. 1650-9.

3.Schomburg, I., et al. Nucleic Acids Res, 2013. **41**(Database issue): p. D764-72.

4.Park, B.H., et al. Glycobiology, 2010. **20**(12): p. 1574-84.

5.North, S.J., et al. J Biol Chem, 2010. **285**(8): p. 5759-75.

6.Liu, G., et al. Bioinformatics, 2008. **24**(23): p. 2740-7.

7.Liu, G., et al. Bioinformatics, 2013. **29**(3): p. 404-6.

(490h) Integrating Omics Data Into Metabolic Models of Lipids: A Platform for the Identification of Gene Mutations From Lipidomics Data

Savoglidis, G., University of Patras
Angelino, P.,
Riezman, H.,
Hatzimanikatis, V., Swiss Federal Institute of Technology of Lausanne (EPFL)
Originally presented on: 11/6/2013 14:36:00 - 14:54:00

Kinetic models of metabolic networks are mapping the biochemical reactions taking part in a cell to a mathematical representation. They are the blueprints of the biochemical network, connecting enzymatic activities and metabolic compounds. Kinetic models have been used in systems biology to produce estimates of response of an organism to stress and to reveal potentially efficient targets for cellular engineering.

Lipids are major constituents of the cell membrane. They are responsible for major properties of the cellular membranes: hydrophobicity, selective permeability and being the scaffold of signaling proteins. Many diseases are associated with alterations in the lipid distribution in the cell and the composition of membrane domains. Metabolic syndrome, obesity, atherosclerosis, as well as Alzheimer's, Huntington's diseases and cancer, all originate from alterations in some stage of lipid biosynthesis. Therefore, advancement of knowledge in the field of lipid metabolism will provide novel insights for further biomedical research and potential strategies for drug development.

In the current study we combine lipidomics data and metabolic models of lipid metabolism in yeast *Saccharomyces cerevisiae*. We use the models as platform for the application of sensitivity analysis and metabolic control analysis, which involves *direct* and *inverse sensitivity analysis*. In *direct sensitivity analysis* we quantify how the changes in enzymatic activities, either by gene deletion or RNAi, are mapped through the metabolic pathway model into variations in lipid profiles. This analysis allows us to identify the effects of the perturbations in enzyme activities (inhibition or overexpression) on the distribution of the lipids synthesized as the end products of the network.

We further extend the analysis by casting the inverse problem. With the *inverse sensitivity analysis* framework, we consider as input an altered lipidomic profile and we are able to identify, the responsible perturbations in enzyme activities that account for such a change. The results of the analysis have been confirmed based on lipidomics analysis of mutant yeast cells.

Although we demonstrate the application of the method in models of lipid metabolism it is possible to be applied to biochemical systems of various parts of metabolism and sizes creating a modular platform for identification of mutations.

(547a) Towards a Whole-Cell Model of Escherichia Coli

Ruggero, N., Stanford University
Macklin, D., Stanford University
Birch, E., Stanford University
Covert, M. W., Stanford University
Originally presented on: 11/6/2013 15:15:00 - 15:33:00

The ability to predict molecular phenotypes at the genome scale is crucial to understanding biological systems. In an effort to address this need we recently published a gene-complete whole-cell model of Mycoplasma genitalium that computationally represents all functions and interactions of every annotated gene [1]. We now aim to extend the whole-cell modeling methodology to the Gram-negative bacteria Escherichia coli. First we will model several core aspects of E. coli physiology including transcription, translation, RNA/protein decay, replication, and metabolism. We will accomplish this by adapting modules of our M. genitalium model [1] to E. coli physiology, and developing new modules to represent E. coli unique phenomena. Furthermore, we will continue to update and streamline the whole-cell modeling infrastructure, improving tools for whole-cell model data curation, construction, simulation, and exploratory analysis. Once finished, the model will enable researchers to investigate single-cell E. coli physiology.

[1] J.R. Karr*, J.C. Sanghvi*, D.N. Macklin, M.V. Gutschow, J.M. Jacobs, B. Bolival, N. Assad-Garcia, J.I. Glass, M.W. Covert. "A Whole-Cell Computational Model Predicts Phenotype from Genotype". Cell. 2012. 150(2): 389-401

(547b) Large-Scale Reconstruction and Analysis of High Quality Core Metabolic Models for Over 8000 Microbial Genomes

Edirisinghe, J. N., Argonne National Laboratory
Conrad, N.,
Frybarger, P.,
Shabbeer, A.,
Henry, C. S., Argonne National Laboratory
Originally presented on: 11/6/2013 15:33:00 - 15:51:00

Hundreds of microbial genomes are now being sequenced every day, demanding the analysis of a wide variety of microbes that may be utilized in multiple medical and industrial applications. Metabolic models are becoming widely popular for the high-throughput prediction of microbial phenotypes and behaviour. The Model SEED framework (Henry, DeJongh et al. 2010) provides a means for automatically constructing genome-scale metabolic models using high quality annotations from SEED subsystems (Aziz, Bartels et al. 2008). One of the common problems identified in draft models generated by the Model SEED is a lack in accuracy of pathways predicted for electron transport and ATP production. Central carbon pathways, electron transport chain, and fermentation pathways play an essential role in accurate metabolic modelling, and these pathways depend on environmental factors such as carbon source/electron donor, presence of oxygen or other anaerobic electron acceptors and the fermentation capability of the organism. Unlike eukaryotic organisms, bacteria electron Transport Chain (ETC) are highly diverse and complex, being categorized into obligate aerobic, obligate anaerobic, photosynthetic and facultative. As a result, the terminal electron acceptors range from organic compounds such as dimethyl sulfoxide (DMSO) to endless list of inorganic compounds such as nitrate, nitrite, sulphate, iron, etc. To overcome this problem, we developed tools to build high-quality core models based on well-studied, phylogenetically diverse model organisms, including *Escherichia coli*, *Bacillus subtilis, Pseudomonas aeroginosa, Clostridium acetobutalicum and Paracococcus denitrificans*. High quality core models may then be integrated into genome scale metabolic models, significantly improving the accuracy of energy metabolism pathways.

We applied these tools to produce core metabolic models for nearly 8000 genomes, designed to be accurate depictions of central carbon metabolism, ETC, and energy production for microbial organisms. Building and analysing core models on almost every single sequenced bacterium helps to understand their metabolism on diverse environments and the metabolic variation in relationship to the phylogenetic distances. The ability of these models to correctly analyse electron transport and fermentation diversity of bacteria in various environmental conditions is critical for many industrial applications such as bioenergy production experiments. Simulating gene knockout experiments to optimize desired fermentation products such as ethanol or butanol, can be performed on a large variety of bacteria for industrial use before starting bench experiments. We also compare these models to explore the variability of these pathways that exist in microbial life, and by analysing the ability of our core models to synthesize ATP in the known growth conditions, we assess the extent to which functional fermentation pathways or electron transport chains are known for the organisms. In this study, we identify gaps in our knowledge of these pathways.

Additionally, all tools developed for reconstruction, visualization, and analysis of core models were built into the DOE Systems Biology Knowledgebase (http://kbase.science.energy.gov), where they are available for use by the scientific community. We will highlight these tools in the context of our core model analysis.

(547c) Constrictor: Flux Balance Analysis Constraint Modification Provides Insight for Design of Biochemical Networks

Erickson, K.,
Chatterjee, A., University of Colorado Boulder
Originally presented on: 11/6/2013 15:51:00 - 16:09:00

The use of *in silico* methods has become standard practice to correlate the structure of a biochemical network to the expression of a desired phenotype. Flux balance analysis (FBA) is one of the most prevalent techniques for modeling cellular metabolism. FBA models range from genome-scale reconstructions to simple minimal reaction sets, and have been successfully applied to obtain predictions of growth, theoretical product yields from heterologous pathways, and location of engineering targets to maximize product yield or design antibiotics. We take inspiration from high-throughput recombineering techniques, which show that combinatorial exploration can reveal optimal mutants, and apply the advantages of computational techniques to analyze these combinations. We introduce Constrictor, an *in silico* tool for flux balance analysis that allows gene mutations to be analyzed in a combinatorial fashion, by applying simulated constraints accounting for up- and down-regulation of gene expression. We apply this algorithm to study ethylene production in *Escherichia coli* through the addition of the heterologous ethylene-forming enzyme from *Pseudomonas syringae*. Targeting individual reactions as well as sets of finite numbers of reactions results in theoretical ethylene yields that are as much as 65% greater than yields calculated using typical FBA. Constrictor is an adaptable technique that can be used to generate and analyze disparate populations of *in silico* mutants, select gene expression levels, and troubleshoot metabolic networks.

(547d) In Silico Model-Based Cofactor Engineering for Strain Improvement

Lakshmanan, M., National University of Singapore
Chung, B., Bioprocessing Technology Institute (A*STAR)
Liu, C.,
Lee, D. Y., National University of Singapore
Originally presented on: 11/6/2013 16:09:00 - 16:27:00

Cofactors such as NAD(H) and NADP(H) provide the redox carriers for several metabolic reactions, both anabolic and catabolic, and serve as mediators in energy transfer across the cells. Thus, establishment of overall redox cofactor balance is considered to be one of the key steps in metabolic engineering for achieving the desired cellular physiology. Of several methods to manipulate the intracellular cofactor regeneration rates, altering the cofactor specificity of a particular enzyme is a promising method. However, the identification of relevant enzyme targets for cofactor engineering (CE) is often very difficult and labor intensive. Therefore, development of more systematic approaches to find the appropriate CE targets using the genome-scale *in silico* metabolic modeling and analysis is highly required. In this regard, herein we propose a novel mathematical framework, cofactor modification analysis (CMA) for the systematic identification of suitable CE targets while considering the global metabolic effects upon cofactor balancing. Subsequently, we demonstrate the applicability of the CMA algorithm by applying it to *E. coli* via its genome-scale metabolic model *i*JO1366, thereby identifying the growth-coupled CE targets for overproducing four of its natural products: ethanol, lactate, formate and acetate, and three non-natural products: lycopene, 1,3-propanediol and 1-butanol. In addition, we also demonstrate the use of *in silico* protein redesign techniques to test the feasibility of altering the cofactor specificity from NAD(H) to NADP(H) or vice-versa before it could be experimentally validated. In general, this study outlines an *in silico* model-driven approach for improving the overall redox balance of host strain to achieve the preferred bioprocess objective.

(547e) Using Hierarchical Modularity to Identify Substrate Cycles in Metabolic Networks

Sridharan, G. V., Tufts University
Ullah, E., Tufts University
Hassoun, S., Tufts University
Lee, K., Tufts University
Originally presented on: 11/6/2013 16:27:00 - 16:45:00

An important motif in metabolic networks is the substrate cycle, which describes a set of reactions that serve to replenish any metabolite participating in the cycle without a net catabolic or anabolic function. These cycles would be thermodynamically infeasible without the net consumption of an energy cofactor or the net transfer of ions across membranes. The most commonly known substrate cycles involve reactions from glycolysis and gluconeogenesis that are simultaneously active: inter-conversion of glucose and glucose 6-phosphate; fructose 6-phosphate and fructose 1,6-bisphosphate; and phosphoenolpyruvate and pyruvate. Once thought to be futile, and hence the synonymous terminology in literature, these cycles have more recently been ascribed roles in cellular functions such as thermogenesis and metabolic control. In this regard, substrate cycles in adipose tissue, namely the esterification and hydrolysis of triglycerides, have been investigated as bioenergetics targets for the treatment of obesity [1]. Differential expression of substrate cycle enzymes has also been explored as a possible approach to alter cancer cell metabolism [2].

With the advent of large-scale model reconstructions of metabolic networks, computational tools can lead to the systematic discovery of novel substrate cycles previously unexplored. Gebauer and coworkers recently pointed that substrate cycles and cyclical elementary flux modes (EFMs) are equivalent, and showed that the latter can be identified by applying existing EFM enumeration algorithms, provided that the model network has been appropriately reduced to remove exchange reactions and associated metabolites [3]. However, complete and exhaustive EFM enumeration on large scale networks is not yet computationally feasible, and the prior work limited the EFM search to only those substrate cycles involving ATP consumption. Indeed, many substrate cycles previously described in literature also involve NADH, NADPH or ion transfer across membranes.

We propose in this study a more targeted approach for the computational identification of potential substrate cycles in the context of the hierarchical modularity of the metabolic network. We apply ShReD-based partitioning [4] (our previously published network partition algorithm based on Shortest Retroactive Distances) on a large scale liver network (hepatonet1) [5], and conduct an exhaustive EFM analysis on individual modules rather than the whole network using EFMtool [6]. Since ShReD-based modules are identified based on cyclical interactions, they are in principle conducive to identifying substrate cycles. The EFM search is exhaustive for all but a small number of modules at the top of the hierarchy, which contained too many reactions for EFMtool to complete the enumeration within a reasonable time. In this manner, the discovered cyclical EFMs, which are possible substrate cycles, are placed in the context of the overall functional organization of the network. We identify cyclical EFMs involved in transport, lipid synthesis, folate metabolism, sugar metabolism, and amino acid metabolism, with the net transformation of several different cofactors including NADH, NADPH, as well as other hub metabolites such as sulfate, phosphate, and hydronium ions.

1. Tseng Y-H, Cypess AM, Kahn CR (2010) Cellular bioenergetics as a target for obesity therapy. Nature reviews Drug discovery 9: 465–482.

2. Locasale JW, Cantley LC (2010) Altered metabolism in cancer. BMC biology 8: 88.

3. Gebauer J, Schuster S, De Figueiredo LF, Kaleta C (2012) Detecting and investigating substrate cycles in a genome-scale human metabolic network. The FEBS journal 279: 3192–3202.

4. Sridharan GV, Hassoun S, Lee K (2011) Identification of Biochemical Network Modules Based on Shortest Retroactive Distances. PLoS Computational Biology 7.

5. Gille C, Bölling C, Hoppe A, Bulik S, Hoffmann S, et al. (2010) HepatoNet1: a comprehensive metabolic reconstruction of the human hepatocyte for the analysis of liver physiology. Molecular systems biology 6: 411.

6. Terzer M, Stelling J (2008) Large-scale computation of elementary flux modes with bit pattern trees. Bioinformatics (Oxford, England) 24: 2229–2235

(547f) Computational Investigation of Photosynthetic Host Strains for Industrial Biotechnology

Mehta, K., Stanford University
Swartz, J. R., Stanford University
Originally presented on: 11/6/2013 16:45:00 - 17:03:00

As photosynthetic organisms attract rapidly increasing attention as host strains for sustainable chemical production, it will be essential to understand fundamental aspects of their metabolism to guide strain design and optimization. We believe computational approaches using reconstructed metabolic models can be powerful tools in these efforts.

In particular, the efficiency of photosynthetic assimilation of light energy and atmospheric CO2 into useful products is currently far too low to be economically feasible. Of the two types of organisms that have been studied the most, cyanobacteria are simpler and easier to engineer, and have been the focus of most published studies, although reported productivities and efficiencies have been low. Eukaryotic algae are more complex and difficult to engineer but can, according to some reports, accumulate higher levels of hydrocarbon products, making them potentially more attractive.

We have used metabolic reconstructions of two photosynthetic organisms - iJN678, of a cyanobacterium, and iRC1080, of a eukaryotic alga - to compare these two main classes of photosynthetic organisms as host strains and shed light on fundamental limitations to high-efficiency chemical production. We hope this work will help inform future engineering projects and contribute to a richer understanding of photosynthetic metabolism.

(547g) Comparative Modeling of Metabolism Across the Shewanella Genus

Ong, W. K., Illinois Institute of Technology
Vu, T. T., University of Wisconsin - Madison
Lovendahl, K., University of Wisconsin-Madison
Llull, J., University of Arizona
Reed, J. L., University of Wisconsin-Madison
Originally presented on: 11/6/2013 17:03:00 - 17:21:00

Shewanella is a genus of facultatively anaerobic, Gram-negative bacteria that have highly adaptable metabolism which allows them to thrive in diverse environments. This quality makes them attractive target bacteria for research in bioremediation, biofuel production, and microbial fuel cell applications. Constraint-based modeling is a useful tool for helping researchers gain insights into the metabolic capabilities of these bacteria. While the genomes for 22 *Shewanella* species have been sequenced, *Shewanella oneidensis* (MR-1) is the only species with a genome-scale metabolic model constructed [1]. In this work, we updated the MR-1 model based on experimental evidence of certain pathways. We also constructed metabolic models for three other species, namely *Shewanella sp.* MR-4, *Shewanella sp.* W3-18-1, and *Shewanella denitrificans* (Sden) which span the genus based on the number of shared genes. We also constructed a core *Shewanella* model which consists only of the genes shared by all 22 sequenced species. Model comparisons between the five constructed models were done at two levels – for wildtype strains under different growth conditions and for knockout mutants under the same growth condition. In the first level, growth/no-growth phenotypes were predicted by the models on various carbon sources and electron acceptors. Cluster analysis of these results revealed that MR-1 is most similar to W3-18-1, followed by MR-4 and Sden when considering growth phenotypes. However, a cluster analysis done based on gene content revealed that MR-1 is more similar to MR-4, followed by W3-18-1 and Sden. As a second level of comparison, we identified differences in reactions and gene content which give rise to different functional predictions of single gene knockout mutants using CONGA [2]. Here, we showed how CONGA can be used to find metabolic reaction and gene differences between models. In conclusion, we have developed four species-specific models and a general core model that can be used to do various *in silico* studies of *Shewanella* metabolism. The developed models provide a platform for a systematic investigation of *Shewanella* metabolism to aid researchers using *Shewanella* in various applications.

1. Pinchuk GE, Hill EA, Geydebrekht OV, De Ingeniis J, Zhang X, et al. (2010) Constraint-Based Model of Shewanella oneidensis MR-1 Metabolism: A Tool for Data Analysis and Hypothesis Generation. PLoS Comput Biol 6(6): e1000822. doi:10.1371/journal.pcbi.1000822

2. Hamilton JJ, Reed JL (2012) Identification of Functional Differences in Metabolic Networks Using Comparative Genomics and Constraint-Based Models. PLoS ONE 7(4): e34670. doi:10.1371/journal.pone.0034670

(547h) Towards Multi-Tissue Type Metabolic Modeling of Maize

Saha, R., The Pennsylvania State University
Simons, M., The Pennsylvania State University
Kumar, A., The Pennsylvania State University
Guillard, L., Institut Jean-Pierre Bourgin
Hirel, B., Institut Jean-Pierre Bourgin
Maranas, C. D., The Pennsylvania State University
Originally presented on: 11/6/2013 17:21:00 - 17:39:00

Zea Mays (commonly known as maize) is an important model C_4 plant due to its widespread use as a cereal and energy crop. In this talk we will highlight our progress to develop multi-tissue type metabolic model of *Zea Mays* metabolism assembled by integrating our earlier model *i*RS1563 with information from the KEGG, MaizeCyc, and MetaCrop databases. By including intracellular and intercellular transport and metabolic reactions, the model is tailored to describe functions of five major cell-types in maize: root, stalk, leaf, tassel and seed. Thus far, we complete reconstructing the leaf model, while development of rest of the cellular models is still in progress. The constructed leaf metabolic model contains GPR associations and elemental and charge balanced reaction entries while incorporating experimentally determined biomass composition, transcriptomic and proteomic data. The earlier *i*RS1563 metabolic model contained gaps in pathways (e.g. sterol biosynthesis, sphingolipid biosynthesis, ubiquinone biosynthesis and starch degradation), limited enzyme localization information, and an approximate representation of the PSI and PSII reactions. In addition to resolving shortcomings of the *i*RS1563 model, this second-generation leaf model includes detailed cell localization based on gene expression data and literature evidence. The model contains over 4,400 reactions involved in both primary and secondary metabolism. As many as 1000 reactions are unique to KEGG, 350 reactions are unique to MaizeCyc, and 40 reactions are unique to MetaCrop. The reactions are distributed in six organelles in the bundle sheath cell (i.e. the mitochondrion, plastid, thylakoid lumen, peroxisome, vacuole, and cytoplasm) and five organelles in the mesophyll cell (i.e. the mitochondrion, plastid, thylakoid lumen, vacuole, and cytoplasm). Using flux balance analysis, we simulated low-nitrogen and high-nitrogen supply conditions and analyzed metabolic flux changes. In addition, reactions underpinning nitrogen utilization efficiency were identified along with the effect of nitrogen supply to biomass yield

(620a) Multiscale Cancer Modeling and Insilico Oncology: Emerging Computational Frontiers in Basic and Translational Cancer Research

Radhakrishnan, R., University of Pennsylvania

Originally presented on: 11/7/2013 8:30:00 - 8:48:00

We describe current advances in multiscale computational modeling and simulation approaches in our laboratory combining models in structural biology and those in systems biology. Our approach combines hybrid and multiscale methods which we have developed with network-based methods in systems biology to enable the construction of quantitative models of signaling networks while retaining the crucial elements of molecular specificity, which are highly relevant to profiling and predictive modeling of clinical mutations in many transformed cell lines. We describe two applications in cell signaling with quantitative and predictive capabilities. (1) To investigate the role of the Anaplastic Lymphoma Kinase (ALK) receptor in pediatric neuroblastoma, we present a computational modeling and simulation approach to delineate molecular-level mechanisms of activation of protein receptor tyrosine kinases and describe clinical implications of the mutations. We show here that our results shed molecular-level insight into the various mechanisms governing such transforming mutations at the level of kinase activity and are remarkably consistent with experimental observations. This study on ALK with enables predictions of driver oncogenic mutations with low false-positive rates, and can hence serve an important in silico tool toward personalized cancer therapy. (2) Somatic mutations in, or over-expression of, the receptor tyrosine kinases of the ErbB family are found in many cancers and is correlated with a poor prognosis. Particularly, clinically identified mutations found in non-small-cell lung cancer (NSCLC) of ErbB1/HER1/EGFR have been shown to increase its basal kinase activity and patients carrying these mutations respond remarkably to the small tyrosine kinase inhibitor gefitinib. Here, we analyze the potential effects of the currently catalogued clinically identified mutations in the ErbB family kinase domains on the molecular mechanisms of kinase activation. Recently, we identified conserved networks of hydrophilic and hydrophobic interactions characteristic to the active and inactive conformation, respectively. Here, we show that the clinically identified mutants influence the kinase activity in distinctive fashion by affecting the characteristic interaction networks. We show that by analyzing the properties of these networks one can gain valuable insight into cohort-sensitivity to inhibition as well as into inhibitor-resistance mechanisms.

Related References:

1. Molecular Systems Biology of ErbB1 Signaling: Bridging the Gap through Multiscale Modeling and High-Performance Computing, Andrew Shih, Jeremy Purvis, R. Radhakrishnan, Molecular Biosystems (A Royal Society of Chemistry Journal), 4: 1151-1159, 2008. Pubmed ID: 19396377

2. Molecular Dynamics Analysis of Conserved Hydrophobic and Hydrophilic Bond Interaction Networks in ErbB Family Kinases, A. Shih, S. E. Telesco, S. H. Choi, M. A. Lemmon, R. Radhakrishnan, Biochemical

Journal, 2011, **436**(2), 241-251. Pubmed ID: 21426301.

3. Analysis of Somatic Mutations in Cancer: Molecular Mechanisms of Activation in the ErbB family of Receptor Tyrosine Kinases, A. J. Shih, S. E. Telesco, R. Radhakrishnan, Cancers, 2011, 3(1), 1195-1231; doi:10.3390/cancers3011195. Pubmed ID: 21701703.

4. Erlotinib binds both inactive and active conformations of the EGFR tyrosine kinase domain, J. H. Park[1], Y. Liu[1], M. A. Lemmon*, R. Radhakrishnan*, Biochem J., 2012, 448(3), 417-423; 1. Equal contribution, * co-corresponding authors.

5. A Multiscale Modeling Approach to Investigate Molecular Mechanisms of Pseudokinase Activation and Drug Resistance in the HER3/ErbB3 Receptor Tyrosine Kinase Signaling Network, S. E. Telesco, A. J. Shih, F. Jia, R. Radhakrishnan, Molecular Biosystems (RSC Journal), 2011, 7 (6), 2066 - 2080. DOI: 10.1039/c0mb00345j. Pubmed ID: 21509365.

6. Computational Methodology for Mechanistic Profiling of Kinase Domain Mutations in Cancers, P. J. Huwe, and R. Radhakrishnan, Proceedings of the IEEE, 5th International Advanced Research Workshop on In Silico Oncology and Cancer Investigation, 2013, in press. An open access version of the article is also available at http://www.5th-iarwisoci.iccs.ntua.gr/

(620b) Lung-Liver Interactions: A Multi-Scale In Vitro Systems Analysis

Young, C. L., University of Delaware
Prantil-Baun, R.,
Ebrahimkhani, M.,
Large, E.,
Hughes, D.,
Borenstein, J.,
Griffith, L. G., Massachusetts Institute of Technology
Lauffenburger, D., Massachusetts Institute of Technology
Originally presented on: 11/7/2013 8:48:00 - 9:06:00

Individual cells integrate many external cues — including those that arise from various extracellular matrix components, mechanical stimulation and soluble signals from adjacent and even distant cells — to generate a basal phenotype and respond to perturbations in their environment [1]. Yet, to capture the complexity of human physiology *in vitro* demands the evaluation of interacting cell types coexisting in a hierarchical 3-dimensional structure influenced by gradients in nutrients, mechanics, and cell composition. Ultimately, to computationally model integrated responses of multiple cells types (including immune cells) requires the need to assess systems under both homeostatic and pathologic conditions. In this study, we propose to integrate communication between lung-liver tissue modules. Specifically, we investigated how changes in the immune cell populations regulate the response to inflammatory signals, such as cytokines and chemokines, and corroborated the effects of cell function and viability across a broad spectrum of metrics.

To investigate the effects of inflammatory stimuli, multi-scale approaches provide a basic framework to assess liver and lung tissues in isolation, as well as in an interacting system. Multivariate analyses of resultant phenotypic responses to external cues were assessed in a dynamic fashion. In this study, we systematically evaluated organ systems crosstalk in the presence of multiple factors, including stimuli that mimic bacterial (e.g. LPS), viral responses (e.g. polyI:C) or cytokines in the presence or absence of drugs to minimize inflammation. Experimentally, standard metrics for cell function were implemented, i.e. retention of Cyp450 enzymes, secretion of serum proteins, gene/protein expression, and immunofluorescence.

Due to the intrinsic complexity of biological systems, integration of experimental and computational approaches is crucial to transform the way we study human tissue physiology and pathophysiology *in vitro*. Here, a systems biology perspective – including the use of statistical techniques such as PCA and PLSDA – led to novel insights attributed to cytokine profiles and signaling networks across a range of physiological perturbations, with emphasis on development and validation of predictive models for cell and tissue behavior relevant in drug discovery.

[1] Griffith, L. and Swartz, M. Capturing complex 3D tissue physiology in vitro. Molecular Cell Biology 2006:211-224.

(620c) Modeling the Complex Interactions Between the Insulin and Angiotensin II Signalling Systems

Arkun, Y., Koc University
Cizmeci, D.,
Originally presented on: 11/7/2013 9:06:00 - 9:24:00

The cross-talk between insulin and Angiotensin II signalling pathways plays a significant role in the co-occurrence of diabetes and hypertension. Systems biology provides a useful framework to study the behavior of complex cellular networks associated with diabetes and hypertension. In [1] Wang developed a mathematical model for the AKT signalling pathway to investigate system-level mechanisms of cell growth and metabolism. The model centers on mTOR which senses the level of nutrients, insulin growth factor which activates the pathway, and pAKT (phosphorylated AKT) that adjusts the growth.

In this work we have advanced Wang's model by including the interactions of AKT signaling pathway with the Angiotensin signalling pathway. The chemical reaction network of the biomolecules involved in the cross-talk is modeled using mass-action kinetics and conservation laws. The resulting model represents a nonlinear dynamical system which is in the form of ordinary differential equations. We show that the complex dynamic intearctions are regulated by a combination of positive and negative feedback loops. These nested feedback loops arc studied to understand their role in important biological mechanisms such as glucose uptake and vasodilation-vasoconstriction balance.

Using this new dynamic model, bistability and parameter sensitivity analysis is carried and the system behavior is mapped to the parameter space to give insight into the development of diabetes and hypertension. The normal regulatory response of insulin signalling is identified to be bistable so that AKT can switch between low and high levels to maintain normal phenotype. The proposed mathematical model predicts that Ang II can impair the insulin signalling pathway through several mechanisms: by direct inhibition of pAKT by ONOO, by directly inhibiting pIRS1 or by activating mTOR. It is shown that the interaction parameters in the model, the balance of positive and negative feedback loops are important in preserving the desired bistability and normal phenotype.

The model delineates that systems with over activated Ang II as a result of disfunction (e.g diseased kidneys can stimulate Ang II) can promote diabetes. The model also predicts that poor glyceamic control induced by diabetes contributes to hypertension by activating the Renin Angiotensin System.

1. Wang G: Singularity analysis of the AKT signaling pathway reveals connections between cancer and metabolic diseases. *Physical Biology* 2010, 7:046015.

(620d) Modeling and Analysis of the Role of Signal Transduction Pathways in the Development of Androgen Independent Prostate Cancer

Rogers, K., Cornell University
Wayman, J. A., Cornell University
Varner, J. D., Cornell University
Originally presented on: 11/7/2013 9:24:00 - 9:42:00

Prostate cancer is the most common cancer in men and the second leading cause of cancer- related death in the United States. Androgens, such as testosterone, are required for prostate cancer growth. Androgen ablation in combination with radiation or chemotherapy remains the primary non-surgical treatment for androgen-dependent prostate cancer. However, androgen ablation typically fails to permanently arrest cancer progression. As testosterone is withdrawn, malfunctioning cells lose their androgen sensitivity and proliferate without hormone. These testosterone insensitive cells often result in Androgen-Independent Prostate Cancer (AIPC). AIPC is closely related to metastasis and decreased survival. In this study, we developed a physiochemical model of an archetypal network, namely, hormone and growth factor signaling in human prostate cancer cells. The model describes the integration of two simultaneous extracellular signaling inputs, namely, androgen and growth factors into a G1/S cell-cycle checkpoint decision (including the regulation of Cyclin D alternative splicing). The model, containing 780 species and 1674 interactions, was identified using data generated from 43 studies in LNCaP cell lines. The model focuses on the outlaw pathway, in which constitutively or overly-activated receptor tyrosine kinases (RTKs) stimulate Akt and mitogen-activated protein kinases (MAPKs), which in turn activate cytosolic androgen receptor (AR) in the absence of androgen. Interestingly, among the few genes activated AR represses is cellular prostatic acid phosphatase (cPAcP), itself a key regulatory of RTK activation. Thus, the outlaw pathway encodes crosstalk between RTK-dependent kinase and AR activation inside of a positive feedback loop. The model was validated with multiple experiments in LNCaP cell lines. In addition data from drug trials was used to evaluate the model's performance predicting the response of prostate tumors to specific drugs. For example, in response to Enzalutamide, an AR inhibitor, the model predicted a decline in PSA expression similar to patient response. Sensitivity analysis, conducted over an ensemble of prostate signaling models, suggested that in an androgen free environment general translation was more sensitive in androgen dependent cells, while in androgen independent cells the PI3K/MAPK pathway species were more sensitive. In a constant androgen environment sensitive species were conserved between the cell lines. These results suggest targeting the PI3K/MAPK pathway in addition to anti androgen therapies as a treatment for AIPC.

(620e) A Quantitative Model of Calcium Regulation in a Homeostatic and Activated Platelet

Dolan, A. T., University of Pennsylvania
Diamond, S. L., University of Pennsylvania
Originally presented on: 11/7/2013 9:42:00 - 10:00:00

A Quantitative Model of Calcium Regulation in a Homeostatic and Activated Platelet

A. Dolan*, S. Diamond

Platelets are the body's first response to vascular damage; upon injury to the endothelium platelets react to the exposed extracellular matrix and undergo a host of intracellular changes enabling them to activate and form a "plug" at the site of injury. The signaling cascades involved in the activation process are complex at both the extracellular and intracellular level. But, as has been seen in many biological systems, calcium plays a critical role, frequently being the focal point of a signaling network diagram. A sharp rise in cytosolic calcium triggers a series of chemical and morphological changes which are critical to platelet activation and clot propagation.

In order to unify the myriad of molecular details of platelet metabolism available in the literature, our lab has developed a compartmentalized model of intracellular calcium signaling in the human platelet. The model consists of a system of coupled ordinary differential equations[1] comprising 84 species, 87 reactions, and 154 rate parameters and spans five well mixed compartments. Calcium balance in the cytosol is determined by the opposing actions of calcium channels and ATPase pumps in the plasma membrane and inner membrane surrounding the platelet's calcium stores. IP3 mediated channels facilitate calcium release from the stores in response to extracellular agonists such as ADP. Store-operated calcium entry (SOCE) channels in the plasma membrane open to allow calcium entry into the cytosol in response to store depletion.

A complete understating of SOCE, a ubiquitous calcium entry pathway seen in many cells types, is still lacking. We have synthesized results from the literature to create a model for SOCE regulation consisting of diffusion-limited dimerization of the calcium sensor STIM1, followed by fast association of STIM1 dimers with trapped Orai channels in the plasma membrane resulting in graded channel activation[2].

Appropriate resting states were characterized using a dense Monte Carlo technique on an initial condition sampling space constrained by physiologic intuition and available species copy number data. Low resting levels of IP3 are required for system robustness and for appropriate dynamic response to physiologic agonists. A ratio of STIM to Orai of six or higher is required for SOCE to have an appreciable contribution to the calcium dynamic during IP3 stimulation and to agree with literature estimates of channel open probability.

References

1. Purvis, J. E.; Chatterjee, M. S.; Brass, L. F.; Diamond, S. L., A molecular signaling model of platelet phosphoinositide and calcium regulation during homeostasis and P2Y1 activation. *Blood* **2008,** *112* (10), 4069-4079.

2. Hoover, P. J.; Lewis, R. S., Stoichiometric requirements for trapping and gating of Ca2+ release-activated Ca2+ (CRAC) channels by stromal interaction molecule 1 (STIM1). *Proc Natl Acad Sci U S A* **2011**.

(620f) Modeling Dorsoventral Patterning of the Drosophila Embryo in Silico Reveals Critical Details Overlooked By Fluorescence Imaging Studies

O'Connell, M. D., North Carolina State University
Reeves, G. T., North Carolina State University
Originally presented on: 11/7/2013 10:00:00 - 10:18:00

For more than 60 years, biologists have been investigating the chemical basis for the establishment of positional information within biological tissues, known as morphogenesis. During morphogenesis, a protein, known as a "morphogen", conveys spatial information by forming a gradient across a tissue through various mechanisms, including local production, diffusion, and global capture. Differential gene expression based on morphogen concentration thresholds initiates tissue-specific responses that lead to cell differentiation and the establishment of distinct body structures. However, recent work modeling gene expression from fluorescence data has raised the question of how tissues far from the morphogen source can be patterned in the face of a low signal-to-noise ratio, causing us to question whether or not measured fluorescence is in fact proportional to the activity of a protein. Here we explore this question from the standpoint of patterning of the dorsal-ventral (DV) axis of the early Drosophila embryo.

The transcription factor Dorsal (dl), a homologue of the mammalian protein NF-kB, controls gene expression along the DV axis of the developing embryo. dl is ubiquitously expressed and sequestered to the cytoplasm by the inhibitor protein Cactus (Cact), which prevents it from promoting or repressing its target genes. However, signaling through the Toll receptor along the ventral side of the embryo causes Cact degradation and nuclear uptake of unbound dl protein in a ventral-to-dorsal gradient. Once dl becomes localized to the nucleus, it can activate or repress its target genes, including *Snail (sna)*, *Short Gastrulation (sog)*, and *Decapentaplegic (dpp)*.

Despite recent developments in imaging techniques that have allowed us to quantify the dynamics of nuclear dl in live embryos, difficulties in modeling gene expression have persisted in regions where the signal-to-noise ratio is low. We hypothesize that this is a result of a misinterpretation of the data gathered with fluorescence imaging studies, which cannot distinguish between the fluorescence of unbound nuclear dl, which regulates gene expression, and nuclear dl-Cact complex, which does not. Previous to this study, it was assumed that dl-Cact complex was found only in the cytoplasm and therefore did not contribute significantly to nuclear fluorescence measurements. However, qualitative disagreement between the more recent experimental data and the established model caused us to question this assumption. By modeling nuclear dl dynamics *in silico*, we were able to separate the two components and model gene expression under the sole influence of unbound dl.

This distinction between active and inactive pools of dl has allowed us to investigate the role of noise in establishing the borders of gene expression seen in the embryo. In fact, we propose that noise may play a significant positive role in determining the slope of gene expression boundaries. Furthermore, we suggest that similar modeling work is needed in other cases in which fluorescence data cannot distinguish between an active form and an inactive form of a biological molecule.

(620g) An Engineering Control System Paradigm for Quantitative Understanding of Hemostasis: A Mathematical Model of the Primary Hemostasis Actuator

Tsai, C. H., University of Delaware
Naik, U. P., University of Delaware
Ogunnaike, B. A., University of Delaware
Originally presented on: 11/7/2013 10:18:00 - 10:36:00

Hemostasis, the physiological process that arrests bleeding and keeps blood within the blood vessel, consists of a set of reactions that can be divided into various functional components. Normal functioning of the components is necessary to achieve a delicate balance between effective hemostasis and pathological conditions such as thrombosis, one of the major causes of death in the world. Because the components of the hemostatic process interact in complex ways, how the various components work together to implement fast, effective and stable responses to vascular injury cannot be fully understood using traditional experimental methods alone—quantitative models are required. While a significant amount of effort has been devoted to developing mathematical models for many individual subsystems of hemostasis in isolation, only a model that adequately captures the complete interactive nature of the subsystems can provide a meaningful representation of the entire process. However, developing a faithful mathematical representation of known mechanistic details of the entire hemostatic process using standard modeling approaches will produce a model containing an inordinately large number of variables and an even larger number of unknown parameters. An alternative conceptual paradigm is required if the important complex details are to be represented adequately, with sufficient fidelity, and in mathematically tractable form.

Our proposed approach is based on the premise that from a process engineering perspective, hemostasis is mediated by an automatic biological control system; an engineering control system representation will therefore provide a structure for organizing information efficiently in terms of the control system components as “functional modules,” thereby facilitating systematic analysis of the complete hemostatic process as a combination of the component modules.

We will present a high-level engineering control system block diagram for hemostasis as the foundational basis of the holistic model and then focus on one subsystem, the process actuator block, which represents how the central physiological process of platelet aggregation is implemented. We will present results showing the dynamic characteristics predicted by our model, compare the model simulations to experimental data, and discuss future plans for completing the model development and using the complete model for systematic analysis of the overall process of hemostasis.

(620h) Elucidating the Effect of Feedback Control in Translation Termination in Yeast

Krishnan, J., Imperial College London

Originally presented on: 11/7/2013 10:36:00 - 10:54:00

While feedback regulation has been widely studied in the context of transcription, feedback regulatory effects in translation have been much less studied. In this talk we will discuss modelling efforts aimed at understanding the feedback control structure involving translation termination in yeast, performed in collaboration with experimentalists. mRNA translation involves roughly three distinct stages, initiation, elongation and termination.Translation resulting in the synthesis of the release factor protein eRF1, involves this protein participating in its own termination. Experimentally introducing an artificial stop codon in the mRNA leads to the possibility of premature termination leading to a truncated protein being synthesized. Thus in this case the release factor negatively regulates its synthesis by its action on the premature stop codon. A very similar feedback regulation in translation is also seen in other systemsIn this talk we will discuss a mathematical modelling framework aimed at elucidating the role of this feedback regulation, and its interaction with tRNA whichmediates premature stop codon readthrough. The translation process is modelled at multiple levels of detail, and the effects of this feedback regulation is systematically elucidated. The dynamical effects of the feedback and its interplay with other steps in the translation process are elucidated using in-silico analysis. We then show that the essential predictions of the in-silico approach are matched by experiments. This work is done in collaboration with Eric de Silva, Russell Betney and Ian Stansfield.

(680a) DATA-Driven Reconstruction Of Biological Networks Using A Nonlinear MODEL Formulation

Asadi, B., University of California, San Diego
Maurya, M. R., Department of Bioengineering, University of California San Diego
Tartakovsky, D. M., University of California, San Diego
Subramaniam, S., University of California San Diego
Originally presented on: 11/7/2013 12:30:00 - 12:48:00

Reconstruction of biological networks is of great interest for learning from large volume of experimental data and facilitating interpretation in biology. There are several methods developed for data-driven reconstruction of biological networks. However, to the best of our knowledge, only a limited number of algorithms have been designed for systematic network reconstruction to account for the nonlinearity of biological systems. Here, we have introduced a novel method to reconstruct nonlinear biological networks. In this method, we use a quadratic nonlinear model as the representation of second-order Taylor series expansion of a nonlinear system around an arbitrary point of interest. As a heuristic, we use a threshold on the correlation between the input/output pairs to identify the informative (significant) inputs in the dataset. We apply LASSO on the significant inputs to shrink some of the small coefficients to zero. Simulation results show considerable improvements in predicting the response of the system and fair improvement in accuracy and sensitivity of the network identified. We have used both synthetic and real biological systems to test our methodology. For the synthetic system, we use different types of nonlinearities to validate the generality of the approach. In the case of real system, we have applied this algorithm to a phosphoprotein (input)/cytokine (output) dataset in RAW 264.7 macrophage cells for which the proposed nonlinear formulation achieves a better performance than a linear model-based formulation of the model.

(680b) Systems Tissue Engineering Approach to Infer Dynamic Transcription Factor Activity Networks Using Large-Scale, Real Time Living Cell Arrays

Penalver Bernabe, B., Northwestern University
Weiss, M. S., Northwestern University
Shin, S., Northwestern University
Asztalos, S., University of Illinois at Chicago
Bellis, A. D., Northwestern University
Tonetti, D., University of Illinois at Chicago
Jeruss, J., Northwestern University
Broadbelt, L. J., Northwestern University
Shea, L. D., Northwestern University
Originally presented on: 11/7/2013 12:48:00 - 13:06:00

Identifying the dynamic activity of critical signaling pathways that lead to a given phenotype and inferring the relationships between those critical signaling pathways are still challenges for the systems biology community. In this work we present a new alternative. Our lab has recently established a technique for large-scale analysis of dynamic signaling pathways within living cell arrays employing lentiviral delivery of transcription factor (TF) reporter constructs, capable of tracking simultaneously the dynamic activities of multiple TFs. We have exploited them to investigate dynamic transcriptional networks of ErbB2 signaling during tissue formation, cancer progression and response to ErbB2-specific therapeutics in 3D hydrogels employing 25 TF reporters in parallel. ErbB2 (HER2/Neu), a member of the EGF receptor family, is overexpressed in a large percentage of aggressive breast cancers, but little is known about its underlying mechanisms or the complete mode of action of the several therapeutic molecules targeting ErbB2 activity.

10A/ErbB2 cells, inducible ErbB2 dimerization MCF10A cells, cultured in Matrigel over 10 days presented different phenotypes and transcription factor activity profiles when treated with EGF or with an ErbB2 dimerization agent (DA). EGF stimulation led to larger spherical structures and generated transient TF reporter activity responses while DA produced larger, highly disorganized structures and targeted more cellular processes stably. DA phenotype required constitutive activation of multiple TFs reporters, such as E2F1, SP1, SRE, STATs and YY1, while EGF effects on TF reporter activity disappeared after 3 days in culture. When DA-activated 10A.ErbB2 cells in 3D culture were treated with lapatinib, larger changes in TF reporter activity were observed at early experimental times in E2F1, ELK1, GATA, P53, RAR and STAT4 reporters, which may be the most direct target processes of lapatinib.

TF networks, represented as a three-level Boolean model, were identified by a combination of protein-DNA interactions (prior knowledge) and an ensemble of inference methods. The most likely active connections (edges) at each experimental time were found by structure optimization through minimization of the difference between the observed data and the response of the Boolean model. Consensus networks showed that cells treated with DA responded with a slow activation of TF reporters up to day 5 in culture, followed by a blossom of TFs that are activated through E2F1, YY1 and STAT3 reporters, with the AP1 reporter being the most important dynamic hub between those days. EGF response was translated sooner through ELK1 reporter to the rest of the cellular processes, with ELK1, NFkB and SP1 being the most important hubs. In the case of lapatinib, its effects were translated through ELK1, GATA, P53, and RAR reporters, with SP1 and STAT4 being the reporters that were responsible for the modulation of the activity of the majority of rest of the significantly altered reporters. After two days in culture, GATA1, P53 and SP1 reporters inhibited E2F1 reporter activity. Based on our computational analysis, we hypothesized that lapatinib might reduce cell proliferation and induce apoptosis through GATA1. In fact, overexpression of GATA1 in DA-activated 10A.ErbB2 cells produced smaller and less disorganized 3D structures and reduced cell viability to similar levels as lapatinib.

Taken together, living cell arrays are a promising tool that can shed light on the most likely processes responsible for the observed phenotype. Combining living cell arrays with dynamic computational approaches can identify novel biomarkers and potential therapeutic targets to control the final cellular phenotype.

(680c) Reconciling Gene Expression Data With Regulatory Network Models â€“ A Stimulon-Based Approach for Integrated Metabolic and Regulatory Modeling of Bacillus Subtilis

Faria, J. P., Argonne National Laboratory
Overbeek, R., Fellowship for Interpretation of Genomes
Taylor, R. C.,
Goelzer, A.,
Fromion, V.,
Rocha, M.,
Rocha, I.,
Henry, C. S., Argonne National Laboratory
Originally presented on: 11/7/2013 13:06:00 - 13:24:00

The reconstruction of genome-scale metabolic models from genome annotations has become a routine practice in Systems Biology research. The potential of metabolic models for predictive biology is widely accepted by the scientific community, but these same models still lack the capability to account for the effect of gene regulation on metabolic activity. Our focus organism, *Bacillus subtilis* is most commonly found in soil, being subject to a wide variety of external environmental conditions. This reinforces the importance of the regulatory mechanisms that allow the bacteria to survive and adapt to such conditions. Existing integrated metabolic regulatory models are currently available for only a small number of well-known organisms (e.g *E. coli* and *B. subtilis*). The *E. coli* integrated model was proposed by Covert *et al* in 2004 and has slowly improved over the years. Goelzer *et al*. introduced the *B. subtilis* integrated model in 2008, covering only the central metabolic pathways. Different strategies were used in the two modeling efforts. The *E. coli* model is defined by a set of Boolean rules (turning genes ON and OFF) accounting mostly for transcription factors, gene interactions, involved metabolites, and some external conditions such as heat shock. The *B. subtilis* model introduces a set of more complex rules and also incorporates sigma factor activity into the modeling abstraction.

Here we propose a genome-scale model for the regulatory network of *B. subtilis*, using a new stimulon-based approach. A stimulon is defined as the set of genes (that can be a part of the same operon(s) and regulon(s)) that respond in the same set of stimuli. The proposed stimulon-based approach allows for the inclusion of more types of regulation in the model. This methodology also abstracts away much of the complexity of regulatory mechanisms by directly connecting the activity of genes to the presence or absence of associated stimuli, a necessity in the many cases where details of regulatory mechanisms are poorly understood.

Our model integrates regulatory network data from the Goelzer *et al* model, in addition to other available literature data. We then reconciled our model against a large set of high-quality gene expression data (tiled microarrays for 104 different conditions). The stimulons in our model were split or extended to improve consistency with our expression data, and the stimuli in our model were adjusted to improve consistency with the conditions of our expression experiments. The reconciliation with gene expression data revealed a significant number of exact or nearly exact matches between the manually curated regulons/stimulons and pure correlation-based regulons. Our reconciliation analysis of the 2011 SubtiWiki regulon release suggested many gene candidates for regulon extension that were subsequently included in the 2013 SubtiWiki update. Our enhanced model also includes an improved coverage of a wide range of different stress conditions.

We then integrated our regulatory model with the latest metabolic reconstruction for *B. subtilis*, the iBsu1103V2 model (Tanaka *et al*. 2012). We applied this integrated metabolic regulatory model to the simulation of all growth phenotype data currently available for *B. subtilis*, demonstrating how the addition of regulatory constraints improved consistency of model predictions with experimentally observed phenotype data. This analysis of growth phenotype data unveiled phenotypes that could only be characterized with the addition of regulatory network constraints.

All tools applied in the reconstruction, simulation, and curation of our new regulatory model are now publicly available as a part of the KBase framework. These tools permit the direct simulation of gene expression data using the regulon model alone, as well as the simulation of phenotypes and growth conditions using an integrated metabolic and regulatory model. We will highlight these new tools in the context of our reconstruction and analysis of the *B. subtilis* regulatory model.

(680d) Identifying Distinct Gene Regulatory Networks Subtending Catecholaminergic Neuronal Subtypes Contributing to Hypertension Development

Park, J., University of Delaware
Ogunnaike, B. A., University of Delaware
Schwaber, J. S., Thomas Jefferson University
Vadigepalli, R., Thomas Jefferson University
Originally presented on: 11/7/2013 13:24:00 - 13:42:00

The pathology of hypertension, a chronic disease affecting nearly 50 million adults in the United States, is multifactorial with many environmental and genetic factors and is not completely understood. However, recent physiological studies indicate that neuronal adaptation is a major component of hypertension pathology. We have obtained preliminary results that support our hypothesis that variable synaptic inputs cause transcriptional changes in neurons controlling blood pressure set point leading to hypertension development. Thus, identifying the gene regulatory networks subtending these neuronal changes driving hypertension development will provide greater insight into this disease pathology.

In order to understand the molecular framework of the observed neuronal changes, we developed gene regulatory network models based on our extension of the constrained fuzzy logic (cFL) approach[1]. The cFL approach converts prior knowledge networks, constrained by literature-based interactions, into refined empirical models that quantitatively describe graded responses of biological relationships using Hill-type functions. The existing published approach considers only binary ON-OFF input values when developing a refined model. However, we have extended this approach by allowing for graded inputs into the network model in order to match the graded input values we observed experimentally *in vivo*. We applied our cFL-based approach to describe quantitatively gene-to-gene interactions underlying neuronal response of individual neurons to graded synaptic inputs. We developed prior knowledge networks and explored multiple topologies within the logic space of the prior knowledge networks using a genetic algorithm. Through this process we identified potential families of models that fit our experimental data similarly well. Furthermore, we identified interactions that were consistently retained or eliminated across the model families. This analysis elucidated which interactions were more relevant in neuronal behavior that was observed experimentally.

Using this approach, we modeled a regulatory network of genes involved in neuronal functions associated with blood pressure regulation, with a specific focus on the angiotensin type 1 receptor (AT1R) mediated signaling pathway. The network model was trained against gene expression data collected from 300 individual catecholaminergic neurons obtained from the brainstem of four rats undergoing a pharmacologically induced acute hypertension challenge. Our initial data analysis revealed six distinct neuronal subtypes that aligned with synaptic input-type. Consequently our modeling approach considered the six corresponding data subsets separately and yielded distinct regulatory networks underlying each neuronal subtype. These distinct network structures suggest that the inputs are modulating, or tuning, the gene regulatory networks governing neuronal behavior. This network tuning is apparent when comparing the refined networks across neuronal subtypes. Specifically, the connectivity and sensitivities between TFs and downstream target genes involved in neurotransmitter production and signaling feedback change across subtypes. These gene regulatory models form the foundation for the development of a tissue-level model that incorporates multiple neuronal subtypes. By integrating multiple neuronal subtypes into an encompassing model we will be able to analyze how these neuronal subtypes affect blood pressure set point definition and regulation.

1. Morris, M. K., Saez-Rodriguez, J., Clarke, D. C., Sorger, P. K., & Lauffenburger, D. a. (2011). Training signaling pathway maps to biochemical data with constrained fuzzy logic: quantitative analysis of liver cell responses to inflammatory stimuli. PLoS computational biology, 7(3)

Research Support: NIH (NIGMS R01 GM083108; NHLBI R01HL111621; NIAAA 5 T32 AA007463-26)

(680e) A Three-Enzyme Cascade During N-Glycan Branching: the Minimal Module for Ultrasensitivity

Liu, G., State University of New York
Neelamegham, S., State University of New York
Originally presented on: 11/7/2013 13:42:00 - 14:00:00

Introduction: Human cell-surface and secreted proteins often bear carbohydrate structures or glycans. Among the glycoforms, protein asparagine linked- (N-) glycosylation is an important type of post-translational modification (PTM). This PTM regulates the pharmacokinetics of biotherapeutic proteins, and also cellular adhesion and signaling properties. Here, N-glycan branching is an important step since it regulates the distribution of hybrid and complex N-glycan structures. Wet-lab experiments and computational models developed by Lau *et al.* [1] reveal that the production of tri- and tetra- antennary N-glycans on cell-surface receptors is ultrasensitive to hexosamine flux. However, the precise components of the reaction network that contribute to ultrasensitivity are yet unidentified.

Materials and Methods: Lau *et al.* [1] describe an initial value problem for N-linked glycosylation with 144 species and 166 reactions. In the current paper, we have systematically reduced this model to yield a three enzyme minimal module that exhibits ultrasensitivity and other functionality that is very similar to the more complex LD2007 model (Figure 1). Such analysis quantifies the concentrations of tri-antennary N-glycans and also other species as a function of hexosamine flux. The source for sensitivity regulation and the role of different model components were quantified using three parameters that fit the Hill equation: i) the Hill coefficient, $\boldsymbol{h}$; ii) the half maximal effective concentration, $\boldsymbol{D_{50}}$ and iii) the maximal response $\boldsymbol{R_m}$.

Results and Discussion: The minimal module for ultrasensitivity in the synthesis of tri-antennary glycans from high mannose structure is presented. It includes a linear three-enzyme cascade (E1-E2-E3), the inter-Golgi transports of glycans, and product recycling steps providing feedback control. Sensitivity analysis was performed on this reduced model by perturbing individual enzymatic rate constants or transport rates by 5% of their nominal values and determining the effect on h, D_{50} and R_m with respect to the final tri-antennary product. Such computational analysis reveals that the enzyme E1 (GnTI) regulates ultrasensitivity steepness (h) and branching activation threshold (D_{50}). E3 (GnTIV) regulates maximum response (R_m). Further, a step-wise decrease of binding affinity for substrate and enzymes in the cascade favors higher ultrasensitivity and lower activation threshold. Finally, the minimal model predicts that fine tuning of the characteristics for ultrasensitivity can be performed by varying enzyme concentration or enzyme binding affinity at various enzyme cascade levels. Such perturbations may regulate N-glycan multiplicity on cell-surface glycoproteins, cell growth and differentiation characteristics.

Conclusions: We identify the minimal components of the N-glycosylation model that impart ultrasensitivity. The fundamental mechanism leading to ultrasensitivity has been studied extensively in systems like the MAPK pathway. These studies demonstrate a significant role for cooperativity, zero-order ultrasensitivity (arising from enzyme cycles operating under saturation regimes) and multisite phosphorylation during MAPK activation [2]. In contrast, our study shows that different mechanisms contribute to ultrasensitivity during N-glycosylation.

References:

1. Lau, K.S., et al. Cell, 2007. **129**(1): p. 123-34.

2. Huang, C.Y., et al. PNAS, 1996. **93**(19): p. 10078-10083.

(680f) Circadian Gating of the Mammalian Cell Cycle Restriction Point: A Mathematical Analysis

Su, J., Georgia Institute of Technology
Henson, M. H., University of Massachusetts Amherst
Originally presented on: 11/7/2013 14:00:00 - 14:18:00

Reliable decision making triggered by dynamic extracellular signals is crucial for proper functioning of mammalian cells. Exploring network motifs of such decision making systems is necessary for understanding fundamental design principles, analyzing system dynamics and revealing possible system vulnerabilities. Control of cell cycle entry under circadian humoral epithelial growth factor (EGF) fluctuations plays a critical role in development, stem cell and progenitor renewal, wound recovery and carcinogenesis. Among the most studied epithelial growth factor receptors (EGFRs), EGF receptor 1 (EGFR1) and human EGF receptor 2 (HER2) initiate the cell cycle, sustain cell survival and contribute to various types of cancers when overexpressed or improperly regulated. Over the last several decades, numerous experimental and modeling studies illustrating EGFR signaling mechanisms and downstream events at the molecular level have been reported. Once thought to be highly entangled, EGFR signaling networks are now known to be spatially and temporally separated into tunable modules.

Humoral variations of EGF levels are sensed by the EGFR system though EGF/EGFR binding. Phosphorylation of EGFR homo- and hetero-dimers occurs within minutes, and the resulting signal rapidly moves forward towards the restriction point (R-point) control machinery. The R-point is underlaid by the bi-stable Rb-E2F switch, a mutual inhibitory lock between the growth-suppressing retinoblastoma protein (Rb) and growth-promoting E2F transcription factors. If the R-point is successfully transitioned, the cell becomes committed to entering the cell cycle and retraction of growth factor has no effect on cell fate. Circadian control of cell cycle entry via EGFRs and other growth factor receptor signaling pathways is crucial for functional coordination of organs and suppression of cancer development. The temporal distribution of cell cycle phases shows strong, tissue specific preferences for specific circadian time windows. Carcinogenesis is often characterized by cell cycle progression gradually losing its synchronization to circadian rhythms. Understanding how circadian signals coordinate cell cycle phases to specific circadian times has potential applications in chronotherapy, chemotherapy and growth factor therapy for cancer treatment. However, *in silico* studies focusing on the effects of external growth factor signals on circadian control of the R-point have not yet been reported.

The goal of this study was to computationally investigate circadian gating of the R-point by coupling an existing molecular model of the core circadian clock to a new, minimal model of the EGFR-ERK signaling system and the R-point switch. Our model was based on the hypothesis that the core clock regulates the global EGF level and coordinates local EGF secretion in different tissues such that the resultant rhythmic EGF fluctuations regulate entry into the cell cycle via EGFR signaling. We established an information flow based minimal model to systematically analyze regulation mechanisms, circadian gating of the R-point under subtle, physiological circadian EGF fluctuations, and the effects of naive (fast and quickly attenuated) and endosomatic (slow and sustained) signaling pathways on system robustness. Our goal was not to develop a detailed molecular model but rather to emphasize essential system elements and to reveal design principles of the growth factor receptor and R-point switch gating machinery. More generally, we systematically studied for the first time a robust decision making system termed the Differential Adaptation Decision Gate (DADG) element. A novel feature of this study was that we combined differential signal adaptation and fast/slow dual signaling pathways with decision gates and switches to reveal fundamental design principles of cell cycle control and circadian gating of cell cycle entry.

(680g) Genome-Wide Conservation Profiling and Size Distribution Analysis of Intergenic Sequences in Prokaryotes Provide Insights to Refine sRNA Searches

Tsai, C. H.,
Liao, R., Univ. of Texas at Austin
Palumbo, M. J., New York State Department of Health
Contreras, L. M., University of Texas at Austin
Originally presented on: 11/7/2013 14:18:00 - 14:36:00

Interest in the search of small RNAs (sRNAs) in bacteria have significantly increased in recent years due to the important regulatory functions that they are now known to be capable of exerting [1]. Development of high-throughput methods (e.g. next-generation sequencing) and more sophisticated computational algorithms has allowed rapid identification of potential sRNA targets in a way that includes complete genome coverage [2-5]. However, given their varying size (50-500nts) and their potential genomic locations in the 5' and 3' untranslated regions as well as in intergenic regions [6], identification and validation of true functionally-active sRNAs have been challenging.

In this study, we analyzed the distribution of profiled sRNAs in 10 different bacteria species using existing transcriptome data. All the intergenic regions were analyzed by WU-BLAST to examine conservation levels relative to other species within or outside their genus. We categorized all members of this bacterial set into different classes according to the conservation patterns that we observed. We will discuss our observations of the correlations between conservation of intergenic regions and sRNA abundance. We also analyzed the size distribution of all intergenic regions in each organism. We will discuss the latter results as well as their implications for refining the discovery and validation of sRNAs in prokaryotes using combined computational and experimental methods.

1. Gottesman S, Storz G, Robertson MP, Joyce GF, Gold L, et al. (2011) Bacterial Small RNA Regulators□: Versatile Roles and Rapidly Evolving Variations. Cold Spring Harb Perspect Biol: ;3:a003798 1.

2. Gelderman G, Contreras LM (2013) Discovery of Posttranscriptional Regulatory RNAs Using Next Generation Sequencing Technologies. In: Alper HS, editor. Systems Metabolic Engineering: Methods and Protocols. Totowa, NJ: Humana Press, Vol. 985. pp. 269-295.

3. Wang Z, Gerstein M, Snyder M (2009) RNA-Seq: a revolutionary tool for transcriptomics. Nat Rev Genet 10: 57-63.

4. Storz G, Vogel J, Wassarman KM (2011) Regulation by small RNAs in bacteria: expanding frontiers. Mol Cell 43: 880-891.

5. Livny J (2012) Bioinformatic Discovery of Bacterial Regulatory RNAs Using SIPHT. In: Keiler KC, editor. Bacterial Regulatory RNA: Methods and Protocols. Totowa, NJ: Humana Press, Vol. 905. pp. 3-14.

6. Kawano M, Reynolds AA, Miranda-Rios J, Storz G (2005) Detection of 5'- and 3'-UTR-derived small RNAs and cis-encoded antisense RNAs in Escherichia coli. Nucleic Acids Res 33: 1040-1050.

(680h) Balancing Antibiotic Resistant Subpopulation in Enterococcus Faecalis Via a Dual Signal Mating-Sensing and Self-Sensing System

Shu, C. C., University of Minnesota
Chatterjee, A., University of Minnesota
Chen, Y.,
Dunny, G. M., University of Minnesota
Ramkrishna, D., Purdue University
Hu, W. S., University of Minnesota
Originally presented on: 11/7/2013 14:36:00 - 14:54:00

Enterococcus faecalis is a pathogen causing frequent hospital acquired infection. It transmits the plasmid encoding antibiotic resistance from donors to plasmid-free recipient cells through conjugation. The process of conjugation and plasmid transfer is regulated by two peptides: iCF10 and cCF10. We uncovered the dual functions of these two peptides recently through a combined experimental discovery and modeling expedition. Intracellularly these two peptides regulate the genetic circuit for conjugative plasmid transfer, while intercellularly they serve as signals of donor and recipient cell concentrations. **iCF10** produced by the donor suppresses conjugation also enables the donor to sense its own concentration; **cCF10** produced by the recipient allows the donor to sense recipient concentration and induces conjugation. By sensing the self-signal iCF10 and mate signal cCF10 the donor titrates its response according to the abundance level of the recipient: a vigorous response when recipient cells are abundant and a more restrained reaction when recipients are in short supplies. Such calibrated response minimizes futile replication of the plasmid and wasteful synthesis of conjugation machinery.

We formulated a mathematical model that spans from molecular regulation of the genetic circuit to the balancing and dynamics of the donor-recipient populations. At the molecular level the model incorporates mechanistic genetic regulation of the initiation of conjugation by iCF10 and cCF10; at the population level, the model considers the dynamics of signal peptides and donor-recipient populations. Various parameters of the model have been previously estimated using reporter genes. The mechanism was cross examined using mutants which were impaired in cCF10 and/or iCF10 production. To further shed light on this dual signaling system the dynamics of transcriptome and proteome were evaluated by RNAseq and iTRAQ proteomics respectively. The genome wide survey enabled us to fine tune the parameter values so that the model truly reflect experimentally observed intracellular variables.

The model was used to prescribe the dynamic behavior of a mixed culture of donor and recipient cells at different abundance levels. Intriguingly the model predicts an incompatibility of a complete conversion of recipient cells to plasmid bearing donors. This was attributed from the suppression of conjugation exerted by the self-sensing signal. We hypothesize that the self-sensing signal plays a donor/recipient balancing role to ensure that the population is endowed with antibiotic resistance while maintaining a subpopulation that is free from the metabolic burden of harboring a particular plasmid. Such a dual signaling system may be rather common in nature. Understanding the balancing act of the signaling system may shed light on new ways of controlling the spread of antibiotic resistance.

AUTHOR INDEX

AUTHOR INDEX

AUTHOR INDEX

American Institute of Chemical Engineers
3 Park Avenue
New York, New York 10016-5991

CURRAN ASSOCIATES INC
proceedings
.com